A GARDENER'S
GUIDE TO
PLANT NAMES

B. J. HEALEY

A GARDENER'S GUIDE TO PLANT NAMES

 An excursion into the mysteries of botanical names; and, I hope, an answer to your friends who fix you with a glassy eye and ask, "What's that in English?"

NEW YORK

CHARLES SCRIBNER'S SONS

Printed in the United States of America
Library of Congress Catalog Card Number 72-1202
SBN 684-12911-6 (cloth)

ACKNOWLEDGEMENTS

Having consulted so many books it seems invidious to
mention only a few but I must acknowledge some of my
sources in *The Royal Horticultural Society Dictionary of Gardening* (London), *Sanders' Encyclopaedia of Gardening*, *Garden Flowers
in Color* (*D. J. Foley; Macmillan; New York*), *Collins Guide to
Bulbs* (*P. M. Synge; Collins; London*), *Dictionary of Botanical Terms*
(*G. F. Zimmer; London*), *The English Rock Garden* (*Reginald Farrer;
Jack; London*), etc.

Above all I must thank my wife for her more than hu-
manly patient typing, checking and retyping, and my old
friend J. S. Vinden for his advice and criticism.

CONTENTS

In memory of Reginald and Emma Herd;
the first a fine gardener himself and the other
who found one of her greatest pleasures in
my own garden during the last years of her life.

Introduction

WHY THE DIFFICULT NAMES?

Gardeners on the whole are amiable people, not unlike fishermen, with whom they often have much in common; the same taste for old clothes, the same quiet patience and occasionally the same propensity for quite startling exaggeration. Like fishermen too—and in a more literal sense—they are down to earth characters who pride themselves on their conservatism, and on their quite reasonable mistrust of anything which smacks of word snobbery or looks like an attempt to air unnecessarily superior knowledge. Many of them, in fact, have a rugged dislike of botanical or scientific names and a persistent belief that the man who uses such is at best showing off, or at worst trying to practise some sort of pedantic one-upmanship. He is likely to be greeted with a sly or pitying grin, and the comment, "I don't know how you do it," and the question, "What's that in English?"

It is a fair query. And the answer is equally simple; that in many if not most cases it is not. Relatively few plants do have English names at all and when we ask for them like that we are merely asking, as I shall show later, for a name which sounds more familiar than two words in botanical language. There are relatively few English names because out of the inexhaustible number of plants which we may now grow in our gardens—and one internationally famous firm of seedsmen lists upward of five thousand merely of the better known—by far the greater number must come naturally from countries foreign to our own. Even if these have, or ever had, common names they would still be incomprehensible to most people except in their native lands, nor could we invent popular names which would be acceptable everywhere. Some kind of order there must be and it is far simpler to adopt a more or less common language and to use a method of classifying and naming all living things according to their similarities and differences. This method is known as taxonomy and although it has its areas of doubt, and sometimes acrimonious debate, it works; it works on the whole very well over most of the world.

Nevertheless let us try the other way at this point. Let us agree to call *Forsythia* "Golden bell flower," as it is sometimes named. If you went to any plantsman who knew his trade and asked him to supply you with a golden bell flower he might at first ask you to be a little more precise, but then if you insisted simply on that he would be perfectly justified in giving you *Campanula thyrsoides* or *Lysimachia punctata*, both of them near weeds, *Clematis tangutica* or *rehderiana*, *Lilium canadense*—if you were extremely lucky—or any one of many other plants, all of which are yellow and several a great deal "bellier" than *Forsythia*. It would have been much more satisfactory to ask for *Forsythia spectabilis* to start with and have done with it. Or again if you asked for bluebells in England you would get what is now known as *Endymion non-scriptus*, in Scotland *Campanula rotundifolia*, the harebell, in Austria *Campanula cochlearifolia* (or *C. pusilla*)—known in some parts of Austria and Germany as "Little blue nun of the meadows"—in Virginia *Mertensia virginica* and in China or Japan, *Platycodon grandiflorum*.

You may say that it is not your intention to go travelling round the world asking for bluebells; but still you can see that even the attempt in that way would be impractical. Clearly it is far simpler to use a standard name of not more than two words, sometimes with a varietal description added, which can be known and recognised everywhere; and without going too deeply into the more obscure botanical classifications, some of which can be highly complex, it is worth while to have at least a working knowledge of these names and what they mean. Nobody can pretend that such knowledge will turn your fingers any greener—a superstition, incidentally, in which I have no belief whatever—nor will it enable you to grow better roses than you do now. But it will certainly add a new dimension of interest to what is at best a dirty, messy, laborious and frequently uncomfortable job. It might inveigle you into looking for some of the more unusual subjects or the lesser

5

known species of those you already grow, to the considerable benefit and variety of your garden; and, although I hesitate to write anything which smacks of fairies in the cabbage patch, it will give plants very definite individualities and characteristics of their own.

"That is all very well," you may argue, "but they are still damned awkward jaw-breaking words," and indeed some of them are; it is almost cruelty, for instance, to saddle one singularly attractive and good humored rock plant with a responsibility like *Chiastophyllum oppositifolium.* But in most cases their apparent difficulty arises only from unfamiliarity. We regularly use a large number of quite correct botanical names already without even noticing we are doing it; and *Calceolaria, Delphinium, Geranium, Antirrhinum, Chrysanthemum,* if you can imagine coming across them for the first time, are by no means easy to pronounce. Yet we all manage it without any difficulty, mainly because we have known them for a long time; we can read them at a glance and we recognise what they mean immediately.

So we admit to knowing *Antirrhinum* and *Geranium* and the rest, but why should we complicate matters by tacking another word onto the end of them; the word to which most people seem to object? The answer is that these are the specific names, the names which separate the individuals from the group just as John is separate from Henry of the Smiths; and given a difference of capital letters, which is a convention, they are written just as you see human names in the electoral rolls, with the surname first—Smith John, Smith Henry, and *Aster alpinus, Aster cordifolius*.[1] To the botanist these names are another classification; to the gardener they are what I mean when I say that looking for some of the lesser known species will be to the benefit and interest of his garden. By putting *asarina* onto *Antirrhinum* you get a trailing plant with yellow snapdragon flowers from June to September for the sunny side of your rock garden, and by adding *armenum* to *Geranium* you have a fine border plant with a constant succession of quite startling magenta-cerise flowers for months on end. Or again by qualifying *Delphinium* with *nudicaule* you find an unusual dwarf species from California with orange scarlet flowers instead of the usual blue, or a yellow one from Persia in *Delphinium zalil*.

There is nothing particularly mysterious or erudite about botanical names—except perhaps sometimes the reasons for giving them—and as every branch of science, profession and trade has its own terminology so are these the terminology of botany and ultimately of gardening. They are no more difficult, while most of them are more interesting and sound a great deal better, than the names of the bits and pieces in your television receiver, and there is certainly no more snobbery in using them than there is in talking about elec-

---

[1] Botanical 'families' in fact are much wider groupings than this; they would correspond more nearly to clans or tribes. But the simplified analogy is enough for our purpose at this point.

trons; a word which is itself based on the same language. Obviously there must be reason in all things, and when a plant has a good, widely recognised English name there is no point in using any other. The man who calls his string beans *Phaseolus coccineus* or his tomatoes *Lycopersicon esculentum*, his tea roses *Rosa hybrida*, deserves all he gets. But surely it is equally foolish to say "The plant named after Achilles which makes you sneeze" when you really mean *Achillea ptarmica*.

HOW IT STARTED AND HOW IT WORKS

In Greek and Roman times, as far back as the fourth century B.C., plants were named by a more or less careful description of their main characteristics, and sometimes from their legendary, medicinal and historical associations. Even with the relatively limited number of plants known in those days it was a cumbersome system, but still it lasted until the fifth century A.D. before it gave way to a shorter and more concise descriptive phrase. This was in Latin as a matter of course, the accepted language of communication, of the Church and of scholarship generally, since the last two through the monasteries and their libraries were closely linked. Again that method lasted until well on to the end of the Middle Ages; and even as late as the sixteenth century, when gardening was becoming a fashionable pursuit for pleasure, John Gerard in the *"Herball, or General Historie of*

Plants" (1597)—of which he lifted a considerable part from an earlier Belgian writer without acknowledgement—still found it necessary to use five Latin words to describe what he called the "Apple Bloome Tulipa."

Clearly the system was again about to become too cumbersome. With the increase of exploration, the growing interest in what were called the Natural Sciences, it could not keep pace with the mood of inquiry and the flood of new, strange and exciting plants; plants like the potato, the South American tomato (originally grown in Europe for decorative purposes) and the wrongly named African marigold which Cortez brought back to Spain from Mexico—all of which Gerard described in the *Herball.* A new simplification was urgently necessary but, although the Swiss botanists John and Caspar Bauhin had devised a method of using only two names for each plant about the year 1600, the true binomial system—the concept of one specific and one generic name for every living organism—only began to appear in the eighteenth century. It was through the work of the Swedish biologist and botanist, Carl von Linné, that this system became widely known and it is now considered to have been first formally set out when he published his *Species Plantarum* in 1753.

Later to be known by the handsome latinised form of his name, Carolus Linnaeus, Carl von Linné was born in 1707. The son of a curate, he became interested in botany as a child when he was called "the little botanist" at the age of eight; 9

he studied at Lund and Uppsala Universities, where he qualified in medicine, and was appointed a lecturer in botany at Uppsala at the age of twenty-three. His first book was published in Amsterdam in 1730, but those which brought him his real fame were *Systema Naturae* (1735), *Genera Plantarum* (1737) and *Species Plantarum* (1753). Thereafter he was made professor of medicine at Uppsala in 1741, and professor of botany in 1742; he died in 1778, when his collection of plants, including more than a hundred new species, was purchased and removed to London. This collection and his library are now the property of the Linnean Society of London, which was founded ten years later "for the cultivation of the science of natural history in all its branches," and was named in his honor.

The career of Linnaeus is not without its humor. "God created, Linnaeus has set in order" his friends announced; but his enemies were not so fullsome, and they described him as "That man who has thrown all botany into confusion." Moreover his romantic imagery of a method of classification according to the arrangement of sexual organs was an affront to the morality of his time, and when he spoke gaily of the reproductive parts of the poppy flower as "Twenty or thirty males in the same bed with a female"—of the stamens and stigma—one of his bitterest critics described the whole Linnean method as "Loathsome harlotry"; and a certain bishop thundered that "Nothing could equal the gross prurience of Linné's mind."

Regardless of criticism, however, he continued to classify unperturbed. There was no living thing that he was not prepared to classify, it was said, from buffaloes to buttercups; and Linnaeus fully agreed, and spent a long, busy and enthusiastic lifetime about it.

In fact classification purely by the number and arrangement of stamens, ovaries and related sexual characteristics did produce some very strange bedfellows in the same families, and it has now been largely superseded. How order, family, genus and species are determined today is extremely complex and still sometimes controversial, but the aim remains to show natural affinities; and the work which is undertaken to this end in the broad field of taxonomy now includes morphology (the study of form and structure), genetics (of origin, reproduction and heredity), cytology (of cells), and serology (the study of serums). But if the method has become more precise the binomial system of Linnaeus has remained, and is likely to remain.

The broad pattern of classification is what can quite reasonably be called the family tree of the plant world. Initially all plants are separated into two primary divisions; those like the fungi, ferns, mosses, algae, etc., which do not have flowers, and the trees, shrubs, plants, grasses and so on which do. The first are known as cryptogams and the second, the flowering plants, as phanerogams. Phanerogams are again subdivided into two more groups determined by the method of carrying their ovules, or the female cells which will ultimately become seed; gymnosperms and angiosperms. Pines, firs, cedars and other cone-bearing plants carrying the ovules exposed are typical gymnosperms, while lilies, daffodils, roses, apples, plums, etc., all with the ovules enclosed or protected in a fruit or capsule, are angiosperms. Then there are still two more simple distinctions, this time based on the development of the seed itself at germination; plants producing one seed leaf being called monocotyledons—such as lilies, crocuses, gladioli—and those which produce two—delphiniums, violas, poppies, primulas and so on—being dicotyledons. (See diagram on pages 21, 22.)

As the techniques of modern botany and taxonomy become more searching there appear to be alternative methods of further classification and there is still some controversy over the way plants are grouped and divided. For our rather simpler purpose it will be enough to pass straight on to what are still sometimes called "natural orders" after Linnaeus, but which now should be more correctly known as "families." This is the last division before genera and species, and it contains groups of plants which are still placed together according to similarities in the arrangement of their various parts; usually in the flower. Family names are recognisable as ending in *-eae* or *-ae* and they will probably be of most interest to the gardener as showing him where his garden plants are related. There is some doubt as to how many of these families there are—the count ranges from 200 to 350 or more according to different schools of botany—but each one includes a varying number of genera, from a few to very many, while each genus embraces in turn a variable number of species; in some cases from only one or two to several hundreds. Thus in the family *Liliaceae* we find the genus *Convallaria* (Lily of the valley) at present with only one species, *majalis* and varieties—which we shall discuss later—*fortunei* and *rosea;* and in the family *Iridaceae* we have *Iris* with several hundred species, varieties and natural and horticultural hybrids, all of widely different characteristics but all recognisable as *Iris* whether they flower in winter or summer, have bulbs, fibrous roots, or rhizomes, and whether they are only a few inches or up to four feet high.

We now have plants of like characteristics in one way or another grouped together in large families. Within these families we have plants bearing different generic names, or surnames, like *Aster, Helenium, Rudbeckia, Cosmos,* all of the family *Compositae;* and each of these surnames will also carry a number, sometimes a large number, of specific or Christian names; every one denoting a distinct individual. Thus we arrive at the Binomial System proposed by Linnaeus; simply that every plant shall be known by two names only although occasionally a third may be added so long as it is qualified by the word "variety" or "var." printed in Roman type. This indicates that the plant so described is undoubtedly of the species stated but varies from it in some noticeable respect. So *Convallaria majalis* var. *rosea* means that the plant is

13

C. majalis but the flower is pink in color, and *Caltha palustris* var. *flore-pleno* indicates that this is *C. palustris* but has double flowers.

First or generic names are a polyglot collection, but always of interest. They may derive from ancient Greek, and sometimes other languages; they may be a form of the name by which the plant was known in its country of origin, or they may honor the gods or heroes of classical Greece or Rome or commemorate some legend. Examples are *Achillea* after a legend of Achilles, *Amaryllis* after a beautiful shepherdess in Theocritus, *Eupatorium* after Mithridates Eupator, King of Pontus, who is said to have discovered an antidote to poison in one of the species, and many more. Often they are based on the name of the person who discovered the genus itself, or honor some well-known figure who need not be connected with horticulture or botany at all. So we shall find *Fuchsia* commemorating Leonard Fuchs, a sixteenth century German botanist, *Clarkia* after Captain William Clark of the Rocky Mountains expedition in the early nineteenth century, and *Saxegothaea* (a tree) in honor of Prince Albert, consort of Queen Victoria and Prince of Saxe-Coburg and Gotha.

Second or specific[1] names are almost equally

[1] Some authorities suggest that the specific name should properly be the complete name of the plant; i.e. the generic and the 'trivial' name or 'epithet' combined. But as this means introducing two more terms here, I have thought it simpler and clearer for our purpose to use the expressions 'generic' and 'specific' names separately.

polyglot but ideally they should be in some way descriptive and, although there are many exceptions, they often are. They can suggest the plant's habit, as *modesta* or unpretentious; its behaviour like *floribunda*, flowering abundantly; where it is found in nature as *nivalis*, near the snow line; its color as *aurea*, golden or orange yellow, and so on. The exceptions are where the specific name refers to the person who first discovered that species, like *farreri, atkinsii, sandersonii*—and for those who like travel books the records of some of the great plant hunters are among the best ever written—where it is named for a woman, like *larpentae* or *olgae*, or where it states the country of which the species is a native; *californica, nepalensis, labradorica, africanus, mexicanum*, and others. These too can sometimes be useful to the gardener, since they may give him an indication of the climatic conditions and surroundings in which the plant lives and presumably thrives.

Finally the initial letter of the generic name is always written with a capital and that of the specific name with a small letter. The exception to this is where an old generic name is now used as a specific epithet after the plant has been reclassified; as an example, *Cneorum* is the ancient Greek name for the shrub which has now been placed in the genus *Daphne* and is called *Daphne Cneorum*. Until recently locational statements like *amurensis* and personal names like *atkinsii* were also shown with capital initials, and sometimes still are, but under recommendations of the International Commission for the Nomenclature of Cultivated Plants (which see later) this practice is now dying out.

THE "FANCY" NAME

In very considerably improving on nature, as there is no doubt the great plant breeders have done—de Graaff of the United States who has developed the lily from a specialist's plant to a magnificent border flower which anyone can grow almost anywhere; Schreiner of Oregon, Fay and others with their work on irises, Reinelt with delphiniums, and many more all over the world—they have also produced some problems of naming and classification. Some of these problems were recognised as long ago as the Botanical Congress of 1866, and of more recent years a code of nomenclature has been drawn up, and is periodically revised, by the International Commission for the Nomenclature of Cultivated Plants of the International Union of Botanical Sciences. This commission represents botanical and horticultural interests from most of the world and its work consists of formulating precise rules for naming cultivars—hybrids, varieties, mutants which have either arisen in cultivation or are maintained there—and for the registration of those names with a national or international authority. Apart from its scientific value this means for the gardener that, when he orders plants or seeds by recognised names, he knows that he will be getting varieties which do have reliably distinct and separate characteristics as described; and for the plant breeder himself—at least in those countries which are signatories to the International Convention for the Protection of New Varieties of Plants, of Paris 1961—it affords a long overdue measure of protection for his rights in cultivars on which he may have spent many years of patient work and development.

Like most systems which seek to cover every possible eventuality the precise rules for naming the various types of cultivar can become somewhat involved and there may be danger in trying to over simplify them; the 1969 Code of Nomenclature, for instance, lays down upwards of fifty articles and recommendations. But for the gardener, at least, such so-called "fancy" names are easily recognisable. With the exception of old varieties which were already known by Latinised names before the modern code was established—like *Iris pseudacorus* 'Variegatus'—cultivar names should not be in Latin, are printed between single quotation marks with capital initial letters, and in Roman type if the preceding botanical name is set in italic; for instance *Delphinium consolida* 'Los Angeles.' It is also permissible, however, when giving the cultivar name to use the generic name only, as *Tagetes* 'Crackerjack,' or even the common name when it is well known; sweet pea 'Ramona.' The naming of hybrids can be more complex—sometimes as complex as the hybrids themselves—but there are three usual methods. For simple hybrids by naming the parents with an X sign between them as *Mimulus cupreus* X *luteus;* by using a name of Latin form after the hybrid has been properly described under the method applying to new species like *Lychnis* X *haageana;* or by writing *Lilium* Bellingham Hybrids when referring to collective hybrids which have emerged as a result of crossing between two species—in this case *L. humboldtii* and *L. pardalinum.*

The International Code does not recognise strains as such, saying that "Any selection showing sufficient differences from the parent cultivar to render it worthy of a name is to be regarded as a distinct cultivar." Most seedsmen do list strains, however, and understandably so, since they represent the end product of a long process of weeding out and reselection to produce plants of consistently high quality which come true to type in form and vigor, and show some particularly desirable garden characteristic. Examples of these are the rust resistant strains of antirrhinum, the Steichen strain of Connecticut Yankees delphiniums, the Barnhaven strains of polyanthus, etc., and they may be regarded, perhaps rather loosely, as the final classification for the gardener.

THE SCOPE OF THIS BOOK

There are now said to be about a quarter million known species of flowering plants in the world. Many of these, if not the greater part, are hardly garden subjects; but even of the number remaining those which can be contained in one book must of necessity be relatively small. The genera and species included here therefore are limited to those plants which may be grown in the open garden in areas of temperate summer and winter conditions, while trees and the larger shrubs are excluded not only for reasons of space but because they are a subject on their own. Species listed have been chosen as for the interest of their specific names, as generally being available through one nursery, bulb, or seedsman or another and, unless otherwise stated, as being good or interesting garden plants.

The book has been planned to lead from the most widely used common names to their genera or species, and then on to a free translation of specific names or epithets. To avoid confusion common names have been given only the first capital initial and placed either in brackets or where necessary in double quotation marks, while horticultural or fancy names are shown by capitals to each word and enclosed in single marks.

Botany is by no means an exact science. There are still areas of doubt and even of controversy, and where these occur I have given what appears to be the most reliable opinion, as I have also in the derivation of some of the more obscure plant names. The aim throughout has been to compile a readable book rather than a directory and, above all, to show that when the chill is taken off it there is a world of interest, information, and sometimes even amusement in the strange and wonderful language known as "Botanical Latin."

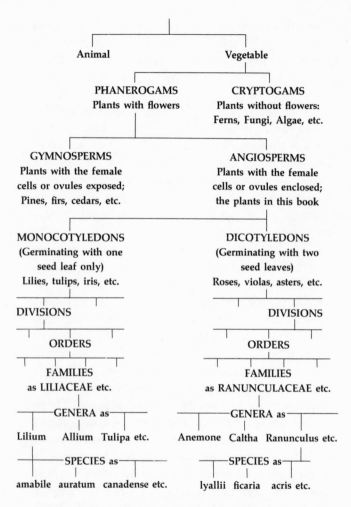

Animal   Vegetable

PHANEROGAMS
Plants with flowers

CRYPTOGAMS
Plants without flowers:
Ferns, Fungi, Algae, etc.

GYMNOSPERMS
Plants with the female
cells or ovules exposed;
Pines, firs, cedars, etc.

ANGIOSPERMS
Plants with the female
cells or ovules enclosed;
the plants in this book

MONOCOTYLEDONS
(Germinating with one
seed leaf only)
Lilies, tulips, iris, etc.

DICOTYLEDONS
(Germinating with two
seed leaves)
Roses, violas, asters, etc.

DIVISIONS   DIVISIONS

ORDERS   ORDERS

FAMILIES
as LILIACEAE etc.

FAMILIES
as RANUNCULACEAE etc.

GENERA as

Lilium   Allium Tulipa etc.

GENERA as

Anemone Caltha Ranunculus etc.

SPECIES as

amabile auratum canadense etc.

SPECIES as

lyallii ficaria acris etc.

A much simplified and generalised diagram to illustrate the descent of the species. True botanical classification is more complex than this, but the various systems used by botanists are still a matter of some controversy.

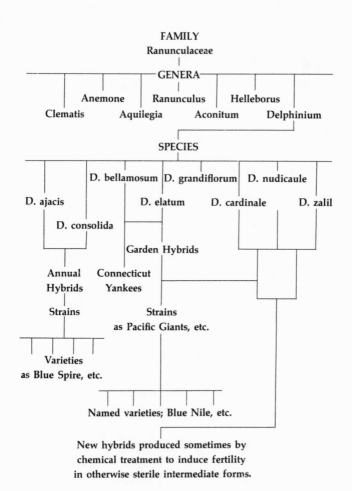

FAMILY
Ranunculaceae

GENERA

Clematis Anemone Aquilegia Ranunculus Aconitum Helleborus Delphinium

SPECIES

D. ajacis D. bellamosum D. grandiflorum D. nudicaule
D. consolida D. elatum D. cardinale D. zalil

Garden Hybrids

Annual   Connecticut
Hybrids   Yankees

Strains   Strains
                       as Pacific Giants, etc.

Varieties
as Blue Spire, etc.

Named varieties; Blue Nile, etc.

New hybrids produced sometimes by
chemical treatment to induce fertility
in otherwise sterile intermediate forms.

An extremely simplified diagram to illustrate the main
lines of development in modern garden delphiniums and
larkspur, and their relationship to some of the other members of the family RANUNCULACEAE.

The Common Names and Their Botanical Equivalents

Where a common name appears to apply to the entire genus or its exact application is obscure no specific name is given.

Aaron's rod	*Verbascum thapsus*
Adam's needle	*Yucca*
African corn lily	*Ixia*
African lily	*Agapanthus*
African marigold	*Tagetes erecta*
Alkanet	*Anchusa*
Alpine wallflower	*Erysimum*
Alum root	*Heuchera americana*
American cowslip	*Dodecatheon*
American Turk's cap lily	*Lilium superbum*
Amerindian lily	*Hymenocallis calathina*
Angel's tears	*Narcissus triandrus*
Apothecary's rose	*Rosa gallica* var. *officinalis*
Apple of Peru	*Nicandra physaloides*
Arum lily	*Zantedeschia aethiopica*
Autumn crocus	*Colchicum*
Avens	*Geum*
Baby blue eyes	*Nemophila insignis*
Baby's breath	*Gypsophila paniculata*
Baby's tears	*Helxine soleirolii*
Balm	*Melissa*

Balloon flower	*Platycodon*
Balsam	*Impatiens balsamina*
Barrenwort	*Epimedium pinnatum*
Basket of gold	*Alyssum saxatile*
Bastard balm	*Melittis melissophyllum*
Bachelor's button	*Centaurea cyanus*
Bear berry	*Arctostaphylos uva-ursi*
Bear's breech	*Acanthus*
Bee balm	*Monarda didyma*
Belladonna lily	*Amaryllis belladonna*
Bell flower	*Campanula*
Bells of Ireland	*Moluccella laevis*
Bergamot	*Monarda fistulosa*
Betony	*Stachys macrantha* (syns. *S. betonica*, etc.)
Bishop's hat	*Epimedium alpinum*
Bishop's wort	*Stachys officinalis*
Birthwort	*Aristolochia*
Black-eyed Susan	*Rudbeckia hirta*
Bladder ketmia	*Hibiscus trionum*
Blanket flower	*Gaillardia aristata*
Blazing star	*Liatris pycnostachya*
Bleeding heart	*Dicentra spectabilis*
Bloodroot	*Sanguinaria canadensis*
Bloody cranesbill	*Geranium sanguineum*
Bluebell (English)	*Endymion non-scriptus*
Bluebell (Spanish)	*Endymion hispanicus*
Bluets	*Houstonia caerulea*
Blue eyed grass	*Sisyrinchium angustifolium*
Blue moonwort	*Soldanella*
Blue poppy	*Meconopsis betonicifolia*
Borage	*Borago*

Bottle gentian	*Gentiana andrewsii*
Bouncing Bet	*Saponaria officinalis*
Buffalo rose	*Callirrhoë involucrata*
Bugbane	*Cimicifuga foetida*
Bugle	*Ajuga*
Bugloss	*Anchusa*
Burnet	*Sanguisorba officinalis*
Burning bush	*Dictamnus*
Bush eschscholtzia	*Hunnemania fumariifolia*
Busy Lizzie	*Impatiens*
Butter and eggs	*Limnanthes douglasii*
Buttercup	*Ranunculus acris*
Butterfly flower	*Schizanthus*
Byzantine squill	*Scilla amoena*
Californian bluebell	*Phacelia minor*
Californian fire cracker	*Brodiaea ida-maia*
Californian fuchsia	*Zauschneria californica*
Californian gum plant	*Grindelia robusta*
Californian pitcher plant	*Darlingtonia californica*
California poppy	*Eschscholtzia californica*
Calla lily	*Zantedeschia aethiopica*
Canary creeper	*Tropaeolum peregrinum*
Candytuft	*Iberis*
Canterbury bell	*Campanula medium*
Cup and saucer type	*C. medium* var. *calycanthema*
Cape hyacinth	*Galtonia candicans*
Cape marigold	*Dimorphotheca aurantiaca*
Caper spurge	*Euphorbia lathyris*
Cardinal climber	*Quamoclit* x *sloteri*
Cardinal flower	*Lobelia cardinalis*
Cardinal monkey flower	*Mimulus cardinalis*

Carnation	*Dianthus* (hybrids)
Catchfly	*Silene*
Catmint	*Nepeta*
Celandine	*Ranunculus ficaria*
Celandine poppy	*Stylophorum diphyllum*
Chamomile	*Anthemis nobilis*
Chapparal yucca	*Hesperoyucca whipplei*
Chequered lily	*Fritillaria meleagris*
Cherry pie	*Heliotropium*
Chilean glory flower	*Eccremocarpus scaber*
China aster	*Callistephus chinensis*
Chinese forget-me-not	*Cynoglossum*
Chinese lantern plant	*Physalis alkekengi*
Chinese pink	*Dianthus* (hybrids)
Christmas rose	*Helleborus niger*
Cinquefoil	*Potentilla*
Clary	*Salvia sclarea or*
	S. horminum
Clover	*Trifolium*
Cobweb houseleek	*Sempervivum arachnoideum*
Cockscomb	*Celosia cristata*
Columbine	*Aquilegia*
Comfrey (common)	*Symphytum officinale*
Comfrey (prickly)	*Symphytum asperum*
Common lavender	*Lavandula spica*
Compass plant	*Silphium laciniatum*
Coneflower	*Echinacea or Rudbeckia*
Coral bells	*Heuchera sanguinea*
Cornflower	*Centaurea cyanus*
Corn marigold	*Chrysanthemum segetum*
Corn poppy	*Papaver rhoeas*
Corsican mint	*Mentha requienii*

Cowslip	*Primula veris*
Cranes bill	*Geranium*
Cream cups	*Platystemon californicus*
Creeping Jenny	*Lysimachia nummularia*
Crown imperial	*Fritillaria imperialis*
Cuckoo pint	*Arum maculatum*
Cup flower	*Nierembergia*
Cup and saucer vine	*Cobaea scandens*
Cupid's dart	*Catananche caerulea*
Cup plant	*Silphium perfoliatum*
Cypress vine	*Quamoclit pennata*
Daffodil	*Narcissus pseudo-narcissus*
Daisy (English)	*Bellis*
Damask rose	*Rosa damascena*
Dame's rocket	*Hesperis matronalis*
Dame's violet	*Hesperis matronalis*
David's harp	*Polygonatum multiflorum*
Day lily	*Hemerocallis*
Devil in a bush	*Nigella damascena*
Devil in the pulpit	*Tradescantia virginiana*
Devil's apple	*Mandragora officinarum*
Devil's bit scabious	*Scabiosa succisa*
Devil's fig	*Argemone mexicana*
Dittany	*Dictamnus*
Dittany (Cretan)	*Origanum dictamnus*
Dog's tooth violet	*Erythronium dens-canis*
Dragon's head	*Dracocephalum*
Drumstick primula	*Primula denticulata*
Dutch lavender	*Lavandula vera*
Dutchman's breeches	*Dicentra cucullaria*
Dutchman's pipe	*Aristolochia macrophylla*
Dyer's chamomile	*Anthemis tinctoria*

Edelweiss	*Leontopodium alpinum*
Elecampane	*Inula helenium*
Evening primrose	*Oenothera biennis*
Everlasting flower	*Ammobium*
	Helichrysum
	Limonium
	Statice
	Xeranthemum
Everlasting or Perennial pea	*Lathyrus latifolius*
Fair maids of February	*Galanthus*
Fair maids of France	*Saxifraga granulata*
Fair maids of Kent	*Ranunculus aconitifolius*
False dragonhead	*Physostegia virginiana*
False hellebore	*Veratrum*
False indigo	*Baptisia*
False spikenard	*Smilacina racemosa*
False sunflower	*Heliopsis helianthoides*
Fireweed	*Epilobium*
Flame flower	*Tropaeolum speciosum*
Flax (common)	*Linum usitatissimum*
Fleabane	*Erigeron*
Floss flower	*Ageratum*
Flower of Adonis	*Adonis*
Flower of a day	*Tradescantia virginiana*
Flower of Jove	*Lychnis flos-jovis*
Flower of the west wind	*Zephyranthes candida*
Foam flower	*Tiarella cordifolia*
Forget-me-not	*Myosotis palustris*
Four o'clock plant	*Mirabilis jalapa*
Foxglove	*Digitalis purpurea*
Foxtail lily	*Eremurus*

French marigold	*Tagetes patula*
Fringed gentian	*Gentiana crinita*
Fumitory	*Corydalis lutea*
Garland flower	*Daphne Cneorum*
Geranium (bedding)	*Pelargonium*
German catchfly	*Lychnis viscaria*
Germander	*Teucrium*
Giant buttercup	*Trollius*
Giant scabious	*Cephalaria*
Giant sunflower	*Helianthus annuus*
Gillyflower (1)	*Cheiranthus cheiri*
Gillyflower (2)	*Dianthus caryophyllus* (hybrids)
Globe flower	*Trollius*
Globe thistle	*Echinops*
Gloriosa daisy	*Rudbeckia tetra*
Glory of the snow	*Chionodoxa luciliae*
Gold band or Golden rayed lily	*Lilium auratum*
Golden flax	*Linum flavum*
Golden rod	*Solidago*
Grape hyacinth	*Muscari*
Grass of Parnassus	*Parnassia palustris*
Guinea-hen flower	*Fritillaria meleagris*
Hag's taper	*Verbascum thapsus*
Harlequin flower	*Sparaxis*
Hawkweed	*Hieraceum*
Heartsease	*Viola tricolor*
Hemp agrimony	*Eupatorium cannabinum*
Heron's bill	*Erodium*
Hoary mullein	*Verbascum pulverulentum*
Hollyhock	*Althaea rosea*

Honesty	*Lunaria*
Hoop petticoat daffodil	*Narcissus bulbocodium*
Hound's tongue	*Cynoglossum*
Houseleek	*Sempervivum*
Humming bird's trumpet	*Zauschneria californica*
Hyssop	*Hyssopus*
Iceland poppy	*Papaver nudicaule*
Immortelle	*Helichrysum*
Indian pink	*Dianthus* (hybrids)
Indian poke	*Veratrum viride*
Indian shot	*Canna indica*
Jacob's ladder	*Polemonium caeruleum*
Jerusalem cowslip	*Pulmonaria officinalis*
Jerusalem cross	*Lychnis chalcedonica*
Joe-Pye weed	*Eupatorium purpureum*
Johnny jump up	*Viola tricolor*
Kaffir lily	*Schizostylis coccinea*
Kansas gayfeather	*Liatris pycnostachya*
Kingcup	*Caltha palustris*
Kingfisher daisy	*Felicia bergeriana*
Lad's love	*Artemisia abrotanum*
Lady tulip	*Tulipa clusiana*
Lady's mantle	*Alchemilla*
Lamb's ear	*Stachys lanata*
Larkspur	*Delphinium ajacis,* *D. consolida* (hybrids)
Lebanon squill	*Puschkinia*
Lemon lily	*Hemerocallis*
Lent lily	*Narcissus pseudo-narcissus*
Lenten rose	*Helleborus orientalis*
Leopard's bane	*Doronicum*

Lily of Peru	*Alstroemeria*
Lily of the Nile	*Zantedeschia aethiopica*
Lily of the Valley	*Convallaria majalis*
London pride	*Saxifraga umbrosa*
Lord Anson's pea	*Lathyrus magellanicus*
Lords and ladies	*Arum maculatum*
Love in a mist	*Nigella damascena*
Love lies bleeding	*Amaranthus caudatus*
Lungwort	*Pulmonaria officinalis*
Lyre flower	*Dicentra spectabilis*
Madonna lily	*Lilium candidum*
Madwort	*Alyssum*
Mallow	*Malva*
Maltese cross	*Lychnis chalcedonica*
Mandrake	*Mandragora officinarum*
Mariposa lily	*Calochortus*
Marjoram (pot)	*Majorana onites* (syn. *Origanum onites*)
Marjoram (wild)	*Origanum vulgare*
Marsh mallow	*Althaea officinalis*
Marsh marigold	*Caltha palustris*
Marvel of Peru	*Mirabilis jalapa*
Mask flower	*Alonsoa*
May apple	*Podophyllum peltatum*
Meadow lily	*Lilium canadense*
Meadow rue	*Thalictrum*
Meadow saffron	*Colchicum autumnale*
Meadow saxifrage	*Saxifraga granulata*
Meadow sweet	*Filipendula*
Mexican fire plant	*Euphorbia heterophylla*
Mexican ivy	*Cobaea scandens*
Mexican marigold	*Tagetes lucida*

Mexican sunflower	*Tithonia*
Mexican tulip poppy	*Hunnemannia fumariifolia*
Michaelmas daisy	*Aster novae-angliae*
	Aster novi-belgii
Mignonette	*Reseda odorata*
Milfoil	*Achillea millefolium*
Milk weed	*Asclepias*
Mind your own business	*Helxine soleirolii*
Missouri sundrop	*Oenothera missouriensis*
Moldavian balm	*Dracocephalum moldavica*
Monkey musk	*Mimulus moschatus*
Monkshood	*Aconitum*
Moonwort	*Lunaria*
Morning Glory	*Ipomoae purpurea*
Moses in the bulrushes	*Tradescantia virginiana*
Moss campion	*Silene acaulis*
Mother of thousands	*Saxifraga stolonifera*
Mountain avens	*Dryas*
Mourning bride	*Scabiosa atro-purpurea*
Mournful widow	*Scabiosa atro-purpurea*
Mourning widow	*Geranium phaeum*
Mullein	*Verbascum*
Mullein pink	*Lychnis coronaria*
Musk mallow	*Malva moschata*
Musk rose	*Rose brunonii*
	(syn. *R. moschata*)
Myrrh	*Myrrhis odorata*
Nailwort	*Paronychia*
Namaqualand daisy (1)	*Dimorphotheca*
Namaqualand daisy (2)	*Venidium fastuosum*
Nasturtium	*Tropaeolum majus*
New Zealand burr	*Acaena*

Night scented stock	*Matthiola bicornis*
Nippon bells	*Shortia uniflora*
Obedience	*Physostegia virginiana*
Oconee bells	*Shortia galacifolia*
Old man	*Artemisia abrotanum*
Old man's pepper	*Achillea millefolium*
Onion	*Allium*
Opium poppy	*Papaver somniferum*
Orange sunflower	*Heliopsis scabra*
Oriental poppy	*Papaver orientale*
Oswego tea	*Monarda didyma*
Ox eye daisy	*Chrysanthemum leucanthemum*
Oxlip	*Primula elatior*
Oyster plant	*Mertensia maritima*
Painted daisy (1)	*Chrysanthemum carinatum* (hybrids)
Painted daisy (2) (Pyrethrum)	*Chrysanthemum coccineum*
Painted tongue	*Salpiglossis sinuata*
Painted trillium or Painted wood lily	*Trillium undulatum*
Pansy	*Viola tricolor* (hybrids)
Partridge berry	*Mitchella repens*
Pasque flower	*Pulsatilla vulgaris*
Periwinkle	*Vinca*
Pheasant's eye	*Adonis annua* (syn. *A. autumnalis*)
Pimpernel	*Anagalis*
Pincushion flower	*Scabiosa atro-purpurea*
Pink	*Dianthus* (hybrids)
Pipe vine	*Aristolochia macrophylla*

Pitcher plant	*Sarracenia*
Plantain lily	*Hosta*
Plume poppy	*Macleaya cordata*
Poor man's orchid	*Schizanthus*
Poppy mallow	*Callirrhoë*
Pot marigold	*Calendula officinalis*
Prickly poppy	*Argemone mexicana*
Prickly thrift	*Acantholimon*
Primrose	*Primula vulgaris*
Prince of Wales feathers	*Celosia plumosa*
Prince's feather	*Amaranthus hypochondriacus*
Prophet flower	*Arnebia echioides*
Puccoon	*Lithospermum canescens*
Purple loosestrife	*Lythrum salicaria*
Purple rock cress	*Aubrieta*
Purslane	*Portulaca oleracea*
Pygmy daffodil	*Narcissus asturiensis* (syn. *N. minimus*)
Pyrethrum	*Chrysanthemum coccineum*
Quamash	*Camassia*
Queen of the prairie	*Filipendula rubra*
Quijote plant	*Hesperoyucca whipplei*
Red hot poker	*Sanguinaria canadensis*
Red puccoon	*Kniphofia (Tritoma)*
Red rose of Lancaster	*Rosa gallica var. officinalis*
Redroot	*Lithospermum canescens*
Red valerian	*Centranthus ruber*
Regal lily	*Lilium regale*
Rhubarb	*Rheum*
Rock cress	*Arabis*
Rockfoil	*Saxifraga*
Rock jasmine	*Androsace*

Rock purslane	*Calandrinia*
Rock rose	*Cistus*
Rose campion	*Lychnis coronaria*
Rose of Heaven	*Lychnis coeli-rosa*
Rosebay willow herb	*Epilobium*
Rose mallow	*Hibiscus moscheutos*
Rosemary	*Rosemarinus*
Rosinwood	*Silphium*
Saffron crocus	*Crocus sativus*
Sage (common)	*Salvia officinalis*
St. Bernard lily	*Anthericum liliago*
St. Bruno's lily	*Paradisea*
St. John's wort	*Hypericum*
St. Patrick's cabbage	*Sempervivum tectorum*
Santa Barbara poppy	*Hunnemannia fumariifolia*
Scarlet sage	*Salvia splendens*
Scots thistle	*Onopordon acanthium*
Sea holly	*Eryngium maritimum*
Sea lavender	*Limonium vulgare* (syn. *Statice limonium*)
Sea pink	*Armeria*
Shasta daisy	*Chrysanthemum* *leucanthemum* (according to Burbank) *or* *Chrysanthemum maximum*
Shell flower (1)	*Chelone obliqua*
Shell flower (2)	*Moluccella laevis*
Shell flower (3)	*Tigridia pavonia*
Shirley poppy	*Papaver rhoeas* (developed from)
Shoofly plant	*Nicandra physaloides*
Shooting star	*Dodecatheon meadia*

Siberian squill	*Scilla sibirica*
Siberian wallflower	*Cheiranthus allionii*
Snake's head iris	*Hermodactylus tuberosus*
Snake's head lily	*Fritillaria meleagris*
Snapdragon	*Antirrhinum majus*
Sneezeweed	*Helenium autumnale*
Sneezewort	*Achillea ptarmica*
Snow in summer	*Cerastium tomentosum*
Snowdrop	*Galanthus*
Soapwort	*Saponaria*
Soldiers and sailors	*Pulmonaria officinalis*
Solomon's seal	*Polygonatum odoratum* (syn. *P. officinale*)
Sops in wine	*Dianthus* (hybrids)
Southernwood	*Artemisia abrotanum*
Sowbread	*Cyclamen*
Spanish bluebell	*Endymion hispanicus*
Speedwell	*Veronica*
Spider flower	*Cleome spinosa*
Spider lily	*Hymenocallis calathina*
Spiderwort	*Tradescantia virginiana*
Spire lily	*Galtonia candicans*
Spotted dog	*Pulmonaria officinalis*
Spring starflower	*Ipheon uniflora*
Star hyacinth	*Scilla amoena*
Star of Bethlehem	*Ornithogalum umbellatum*
Star of Texas	*Xanthisma texanum*
Star of the Veldt	*Dimorphotheca*
Star tulip	*Calochortus*
Statice	*Limonium*
Stock	*Matthiola incana*
(Ten Week Trisomic etc.)	(developed from)

Stonecress	*Aethionema*
Stonecrop	*Sedum acre*
Strawflower	*Helichrysum bracteatum*
Striped squill	*Puschkinia*
Stud flower	*Helonias bullata*
Summer cypress	*Kochia scoparia*
Summer hyacinth	*Galtonia candicans*
Summer snowflake	*Leucojum aestivum*
Sundrops	*Oenothera fruticosa*
Sunflower	*Helianthus*
Sun rose	*Helianthemum*
Swamp pink	*Helonias bullata*
Swan River daisy	*Brachycome iberidifolia*
Sweet alyssum	*Lobularia maritima*
(Sweet Alison)	(syn. *Alyssum maritimum*)
Sweet basil	*Ocimum basilicum*
Sweet Cicely	*Myrrhis odorata*
Sweet pea	*Lathyrus odoratus*
	(hybrids)
Sweet rocket	*Hesperis matronalis*
Sweet scabious	*Scabiosa atropurpurea*
Sweet sultan	*Centaurea moschata*
Sweet William	*Dianthus barbatus*
Tangier pea	*Lathyrus tingitanus*
Tansy	*Tanacetum vulgare*
Tarragon	*Artemisia dracunculus*
Thrift	*Armeria*
Thyme	*Thymus*
Tickseed	*Coreopsis*
Tidy tips	*Layia elegans*
Tiger flower	*Tigridia pavonia*
Toadflax	*Linaria*

Toad lily	*Tricyrtis*
Toad trillium	*Trillium sessile*
Tobacco plant	*Nicotiana*
Torch lily	*Kniphofia (Tritoma)*
Touch me not	*Impatiens balsamina or I. noli-tangere*
Treasure flower	*Gazania*
Tree mallow	*Lavatera arborea and L. trimestris*
Trout lily	*Erythronium revolutum*
Trumpet flower	*Bignonia*
Tuberose	*Polianthes tuberosa*
Tulip poppy	*Papaver glaucum*
Venus's looking glass	*Specularia speculum*
Venus's navelwort	*Omphalodes linifolia*
Vervain	*Verbena officinalis*
Violet (sweet)	*Viola odorata*
Violet cress	*Ionopsidium acaule*
Viper's bugloss	*Echium*
Virginia bluebell (in America)	*Mertensia virginica*
Virginian cowslip (in England)	*Mertensia virginica*
Virginian stock	*Malcomia maritima*
Wake robin	*Trillium grandiflorum*
Wall cress	*Arabis*
Wallflower	*Cheiranthus cheiri (hybrids) and Erysimum*
Wand flower	*Dierama*
Wand plant	*Galax aphylla*
Welsh poppy	*Meconopsis cambrica*

37

Western tiger lily	*Lilium pardalinum*
White campion	*Lychnis alba*
White hellebore	*Veratrum album*
White mayweed	*Matricaria maritima* (syn. *M. inodora*)
White snakeroot	*Eupatorium rugosum*
Willow gentian	*Gentiana asclepiadea*
Windflower	*Anemone*
Winter aconite	*Eranthis*
Wolfsbane	*Aconitum lycoctonum*
Wood betony	*Stachys officinalis*
Wood lily	*Trillium*
Wood sorrel	*Oxalis acetosella*
Wormwood	*Artemisia absinthium*
Yarrow	*Achillea*
Yellow loosestrife	*Lysimachia vulgaris*
Yellow ox-eye	*Buphthalmum*
Zephyr lily	*Zephyranthes candida*

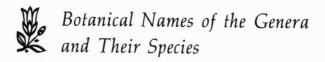

Botanical Names of the Genera and Their Species

Where a common name exists which applies loosely to the entire genus it is shown in brackets immediately following the genetic name; otherwise where it applies only to a single species it appears in that context. The abbreviation "syn" (synonym) indicates an alternative name; due to changes in botanical classification many plants have been given different names over the years, and the current names followed by those superseded are shown as follows: *Iris unguicularis* (syn. *I. stylosa*) or *Chrysanthemum argenteum* (syns. *Matricaria argentea, Pyrethrum argenteum, Tanacetum argenteum*). The color notes refer to flower color except where otherwise stated.

ABELIA

FAMILY: *Caprifoliaceae*

Takes the name of Dr. Clark Abel, a physician and writer attached to the Amherst diplomatic mission to China in 1816–17. Flowering evergreen and deciduous shrubs; not quite hardy in cold winter areas. The following are the most useful where space is limited.

A. chinensis (China; white flushed pink; deciduous), *A. floribunda* (Mexico; rosy red; evergreen), *A* x *grandiflora* (thought to be a hybrid; pink or white; evergreen), *A. schumanii* (China; rosy lilac; evergreen).

ACAENA   (New Zealand burr)

FAMILY: *Rosaceae*

From the Greek *akaina*, a thorn, in reference to the small burrs. A few species of evergreen creeping plants from New Zealand; useful for the rock garden and pavements, although they can become invasive. Perennials.

A. adscendens (bronze foliage, purplish burrs), *A. buchananii* (silvery grey and red), *A. microphylla* (bronze green and crimson), *A. novae-zealandiae* (bronze and brownish red).

ACANTHOLIMON   (Prickly Thrift)

FAMILY: *Plumbaginaceae*

From the Greek *akanthos*, a spine, and *limon*, sea lavender. Small evergreen perennials for the rock garden from Eastern Europe. All have pink flowers and greyish to green foliage. Perennials.

A. acerosum, A. glumaceum, A. venustum.

ACANTHUS   (Bear's breech)

FAMILY: *Acanthaceae*

From the Greek *akanthos*, a spine. Handsome perennials, with statuesque flower spikes lasting for many weeks. Probably the plant most celebrated in architecture since the Greeks adapted its leaf form for the well known decoration on the caps of their Corinthian columns. There are three species most often cultivated, all perennials.

40   *A. longifolius* (purple), *A. mollis* (white or pink)

and variety *latifolius,* with wider leaves, *A. spinosus* (purple and white).

ACHILLEA   (Yarrow)

FAMILY: *Compositae*

Named after Achilles, who is said to have used it to heal wounds at the Siege of Troy, having been taught the uses and virtues of plants by Chiron the Centaur. It is also called the *herbe aux charpentiers* by the French in reference to its use for healing cuts made by carpenters' tools. A large genus widely distributed throughout Europe, the following being among the best for the border and rock garden. Perennials.

For the border: *A. filipendulina* (yellow) and cultivars 'Gold Plate,' 'Canary Bird,' etc., *A. millefolium* (Milfoil, Old man's pepper; white) variety *rosea* and cultivars 'Cerise Queen,' 'Crimson Beauty,' etc., *A. ptarmica* (Sneezewort; white), cultivar 'The Pearl,' *A sibirica* (white), cultivar 'Perry's White.'

For the rock garden: *A. chrysocoma* (yellow), *A. pritchardii* (white), *A. rupestris* (white), *A. tomentosa* (golden yellow).

ACONITUM   (Monkshood)

FAMILY: *Ranunculaceae*

An ancient classical name, either from *aconitum,* a poisonous plant, or the Greek *akon,* a dart, since the juices of the plant are said to have been used at one time as arrow poison. Widely distributed. Handsome tall perennials for the border but all are poisonous, *A. napellus* extremely so, if eaten or in contact with open cuts or scratches.

A. anglicum (Britain; blue), *A. carmichaelii* (China; purple blue), *A. lycoctonum* (Wolfsbane; Siberia; yellow), *A. napellus* (Europe; blue) with varieties *album* (white) and *roseum* (rose) and cultivars 'Bressingham Spire,' 'Spark's Variety,' etc., *A. uncinatum* (syn. *A. volubile;* partly climbing; North America; dark blue), *A. variegatum* (Europe; blue and white).

ADONIS   (Flower of Adonis)

FAMILY: *Ranunculaceae*

Named after Adonis, the beloved of Venus, in a legend that the flower sprang from his blood after he was killed by a wild boar. Other versions give this plant as the rose or the anemone and the myth is probably symbolic of the course of vegetable life; to die in the autumn and reappear in the spring. Annuals and perennials for the front border and rock garden.

Perennials; for the rock garden: *A. amurensis* (China and Siberia; yellow), *A. vernalis* (Europe; yellow).

Annuals; for the border: *A. aestivalis* (Europe; crimson), *A. annua* (syn. *A. autumnalis;* Pheasant's eye; Britain; scarlet).

AETHIONEMA   (Stonecress)

FAMILY: *Cruciferae*

Origin obscure but said to be from the Greek *aitho,* to burn, perhaps a reference to the color; or from the burning taste of some species. Useful trailing and evergreen perennials for the rock garden.

A. armenum (Armenia, now an area of Southwest Asia; blue grey foliage, rose flowers), *A. coridifolium* (Lebanon; distinct blue grey and pink) and cultivar 'Warley Rose,' *A. grandiflorum* (Persia; variable pink), *A. iberideum* (Levantine Alps; white to lilac), *A. pulchellum* (Armenia; rosy purple), etc.

AGAPANTHUS   (African lily)

Family: *Liliaceae*

From the Greek *agape*, love, and *anthos*, flower. Showy perennials from South Africa, most useful for terrace or patio decoration in tubs or large pots. May need some winter protection in colder areas although the modern hybrids are hardier than the species.

A. africanus (blue) with varieties *albus* and *nanus*, *A. campanulatus* (syn. *A. umbellatus*; pale blue) and variety *albus*, *A. orientalis* (variable blue) and 'Headbourne Hybrids' (pale to deep violet blue). All are sometimes erroneously described as *A. umbellatus*.

AGERATUM   (Floss flower)

Family: *Compositae*

From the Greek *a*, not, *geras*, old; referring to the flowers, which do not fade quickly. A popular small annual from Mexico, usually treated as half hardy in less favorable climates. One species and numerous cultivars.

A. houstonianum (syn. *A. mexicanum*); cultivars and varieties 'Florist's Blue,' 'Little Dorrit,' 'Mexican White,' 'Fairy Pink,' etc.

43

AJUGA   (Bugle)

FAMILY: *Labiatae*

From the old Latin name *abiga*. Perennials of creeping habit with upright flower spikes. Two species from Europe with varieties.

A. genevensis (blue), variety *brockbankii* (deep blue), *A. reptans* (blue, white or rose), varieties *atro-purpurea* (dark purple), *multicolor* and *variegata* (blue with variegated foliage), *A. reptans* can become invasive.

ALCHEMILLA   (Lady's mantle)

FAMILY: *Rosaceae*

From an Arabic word *alkemelych,* referring to its one time use in alchemy. Hardy perennials for the rock garden or front border. Several species but only two usually seen in cultivation; both come from Europe and have greenish yellow flowers.

For the front border: *A. mollis.*

For the rock garden: *A. alpina.*

ALLIUM   (Onion; flowering types)

FAMILY: *Liliaceae*

From the Celtic *all,* meaning pungent; or simply the old Latin name for garlic, *allium.* A large and widely distributed genus of bulbous plants mostly with spherical flower heads. Ranging from a few inches to four feet high, suitable species can be used for the wild garden, rock garden, border or as spot plants. Some are valued for formal floral decoration. The following selection is limited; see bulbsmen's lists.

For the border: *A. azureum* (Siberia, Central Asia; blue), *A. giganteum* (Central Asia; lilac), *A. pulchel-*

lum (Southern Europe; lilac purple), *A. ramosum* (Central Asia; white and rose), *A. sphaerocephalum* (Europe; purple), etc.

For the rock garden: *A. breweri* (California; violet purple), *A. narcissiflorum* (European Alps; red purple), *A. ostrowskianum* (Caucasus; rose), *A. triquetrum* (Southern Europe; white), etc.

For the wild garden: *A. moly* (Europe; bright yellow; tends to become invasive).

ALONSOA   (Mask flower)

FAMILY: *Scrophulariaceae*

Named after a Spanish government officer, Alonzo Zanoni, sometimes given as Zanoni Alonzo. Shrubby perennials, half hardy in most areas and usually treated as annuals. A few species in cultivation.

A. acutifolia (Mexico; scarlet) and variety *alba* (white), *A. incisifolia* (Peru; scarlet), *A. linearis* (Peru; scarlet) and variety *gracilis,* etc.

ALSTROEMERIA   (Lily of Peru)

FAMILY: *Amaryllidaceae*

Named after Baron Clas Alstroemer, a Swedish botanist of the eighteenth century. Popular perennials, but liable to injury by severe cold.

A. aurantiaca (Chile; orange to carmine) and variety *lutea* (yellow), *A. brasiliensis* (Brazil; yellow to red or brown), *A. campaniflora* (Brazil; green), *A. chilensis* (Chile; orange), *A. haemantha* (Chile; red to purple), *A. violacea* (Chile; violet mauve) and 'Ligtu Hybrids' (lilac to pink and purple), 'Dr. Slater's Hybrids' (variable colors), 'Moorheim Orange,' etc.

ALTHAEA

FAMILY: *Malvaceae*

From the Greek *althaea*, to cure, referring to the one time medicinal value of these plants. Well known tall annuals, biennials and perennials, of which the species have been largely superseded by the much improved hybrids and cultivars.

Perennials: *A. narbonensis* (Narbonne, France; red), *A. officinalis* (Marsh Mallow; Britain; rose).

Biennials: *A. ficifolia* (Siberia; mostly yellow), *A. rosea* (Hollyhock; China; shades of pink) and cultivars 'Chater's Improved,' 'Begonia Flowered,' etc.

Annual cultivars: 'Annual Double,' 'Triumph Supreme,' etc.

ALYSSUM   (Madwort)

FAMILY: *Cruciferae*

From the Greek *a*, not, and *lyssa*, madness; once supposed to cure rage and the bite of a mad dog. Dwarf perennials for the rock garden; all from Europe and the Mediterranean areas. (For Sweet alyssum, or more correctly Sweet Alison, see *Lobularia*).

A. alpestre (yellow), *A. argenteum* (silvery foliage; yellow), *A. montanum* (bright yellow), *A. pyrenaicum* (silvery foliage; white), *A. saxatile* (Basket of gold) with varieties *citrinum* (lemon yellow), *compactum* (lemon yellow) and cultivars 'Dudley Neville,' 'Tom Thumb,' etc.

AMARANTHUS

FAMILY: *Amaranthaceae*

From the Greek *a*, not, and *maraino*, to fade; referring to the lasting quality of the flowers. Annuals from India and the tropics. Those grown for their brilliantly colored foliage are best as greenhouse plants; the flowering varieties listed here are treated as half hardy annuals in most areas.

A. caudatus (Love lies bleeding; red) with varieties *albiflorus* (greenish white), *atro-purpureus* (dark purple) and *viridis* (green), *A. hypochondriacus* (Prince's feather; deep crimson) with varieties *atropurpureus*, *sanguineus* (blood red) and *splendens* (crimson foliage).

AMARYLLIS   (Belladonna lily)

FAMILY: *Amaryllidaceae*

Named after a beautiful shepherdess in the poems of Theocritus and Virgil. A late flowering bulbous plant from South Africa. American authorities consider the name Amaryllis should correctly be applied to the genus *Hippeastrum*, this plant being known to them as *Callicore rosea* or *Brunsvigia rosea*. The common name is literally 'pretty lady' since an extract from the plant was at one time used by ladies to brighten their eyes.

A. belladonna (syns. *Callicore rosea, Brunsvigia rosea;* generally pink, but white to lilac purple forms are known).

Other genera which are sometimes called *Amaryllis* include *Crinum, Lycoris, Nerine, Sprekelia, Sternbergia, Vallota* and *Zephyranthes.*

AMELLUS

FAMILY: *Compositae*

From the name given by Virgil to a similar flower growing by the River Mella. A small genus

47

from South Africa; only one annual species usually found in cultivation.

A. annuus (syn. *Kaulfussia amelloides;* blue with a yellow center).

AMMOBIUM

FAMILY: *Compositae*

From the Greek *ammos,* sand, and *bio,* to live; describing its habitat. Only one annual species from Australia available, generally treated as half hardy and useful as a dried or everlasting flower.

A. alatum (silvery white and yellow) and variety *grandiflorum.*

ANAGALLIS   (Pimpernel)

FAMILY: *Primulaceae*

From the Greek *anagelao,* delightful; or possibly from a one time belief that an infusion from this plant was a cure for melancholy. Some species are perennial but all are usually treated as annuals. Useful for the front border.

A. arvensis (Europe; white to scarlet), varieties *caerulea* (blue) and *phoenicea* (red), *A. fruticosa* (Morocco; vermilion), *A. indica* (India; blue), *A. linifolia* (Europe; blue) and varieties *breweri* (red), *lilacina* (lilac) and *phillipsii* (gentian blue), *A. tenella* (Britain; pink).

ANCHUSA   (Alkanet; Bugloss)

FAMILY: *Boraginaceae*

From the Greek *anchousa,* a cosmetic paint or stain; possibly a coloring from the blue flowers used by the ancient Greek ladies for eye shadow, although a red infusion may also be prepared from the roots and Gerard remarks in the *Herball*, "The Gentelwomen of France do paint their faces with these roots as it is said." A useful genus since it also provides showy biennials and perennials for the border.

Perennials: *A. azurea* (syn. *A. italica*; Mediterranean regions; deep blue) and cultivars 'Dropmore,' 'Loddon Royalist,' 'Sutton's Royal Blue,' etc., *A. caespitosa* (Crete; gentian blue), *A. sempervirens* (Europe; blue).

Biennials: *A. capensis* (South Africa; blue) and cultivar 'Blue Bird,' *A. officinalis* (Europe; deep blue, sometimes purple), and variety *incarnata* (pink).

ANDROSACE   (Rock Jasmine)

FAMILY: *Primulaceae*

Somewhat fancifully from the Greek *aner*, a man, and *sakos*, buckler; the anther being supposed to resemble an old type of buckler. Small perennial rock plants, most of which are difficult to grow. These following are comparatively easy.

A. alpina (European Alps; pink), *A. carnea* (Europe; pink), *A. chamaejasme* (North America; white), with variety *ciliata* (rose pink), *A. lanuginosa* (Himalayas; rose), *A. sarmentosa* (Himalayas; rose), etc.

ANEMONE   (Windflower)

FAMILY: *Ranunculaceae*

From the Greek *anemos*, wind, and *mone*, habitation, from the fact that some species are found in windy places. A large, variable and widely distributed genus which supplies plants for almost every situation in the garden. Some are now listed separately under *Hepatica* and *Pulsatilla.* The Japanese anemones are hybrids of *A. hupehensis* and *A. vitifolia.* A selection only here; see nurserymen's catalogs.

For the border: *A. canadensis* (North America; white), *A. glauciifolia* (Western China; bluish lilac), *A. hupehensis* (China; rose) variety *japonica* (Japan; deeper carmine) and cultivars 'Bressingham Charm,' 'September Glow,' 'Krimhilde,' etc., *A. rivularis* (Himalayas; white), etc.

For the rock garden: *A. apennina* (Europe; blue, rose and white), *A. blanda* (Asia Minor; blue), *A. fulgens* (Southern Europe; scarlet), *A. magellanica* (South America; pale yellow), *A. nemorosa* (Britain; white), etc.

For cutting: *A. coronaria* (Southern Europe; variable colors) with strains and cultivars 'de Caen,' 'His Excellency,' 'St. Brigid,' etc.

ANTHEMIS

FAMILY: *Compositae*

From the Greek *anthemon,* flower; referring to their profuse blooming. Aromatic perennials for the border or rock garden. Chamomile tea is made from *A. nobilis* and a non-flowering variety of this species is sometimes used for making lawns; according to an old gardening tradition it is said also to be a 'doctor' to other plants and to improve their health when grown close to them.

50 For the border: *A. sancti-johannis* (Bulgaria; or-

ange), *A. tinctoria* (Dyer's chamomile; Europe; yellow) and cultivars 'Beauty of Grallagh,' 'Grallagh Gold,' 'Wargrave Variety,' etc.

For the rock garden: *A. biebersteiniana* (Asia; silvery leaves, yellow flowers), *A. nobilis* (Chamomile; Europe; white) and variety 'Treneague' (nonflowering), *A. rudolphiana* (Europe; golden orange), etc.

ANTHERICUM   (St. Bernard lily)

FAMILY: *Liliaceae*

From the Greek *anthos,* flower, and *kerkos,* a hedge; probably alluding to the fact that this species spreads by means of underground stems which form new plants around the parent and may thus make a hedge. Graceful lily-like plants from Southern Europe. All are white.

A. liliago and variety *major, A. ramosum.*

ANTIRRHINUM   (Snapdragon)

FAMILY: *Scrophulariaceae*

From the Greek *anti,* like, and *rhinos,* snout; describing the curious shape of the flowers; nurserymen now, however, are producing penstemon and azalea flowered types which are quite unlike this typical shape. They are all strictly perennials, but *A. majus,* and the inumerable modern cultivars which are derived from it, are usually treated as annuals. There are two little known trailing species useful for the rock garden. All from Southern Europe.

For the border and bedding: cultivars in variety   51

of *A. majus* (originally pink), see seedsmen's catalogs.

For the rock garden: *A. asarina* (yellow), *A. glutinosum* (cream and yellow).

AQUILEGIA   (Columbine)

FAMILY: *Ranunculaceae*

Probably from the Latin *aquila;* referring to the spurred form of the flower, like an eagle's claw. The common name, Columbine, is said to come from the Latin *columba,* a dove; the form of the flowers suggesting, rather fancifully, a flight of doves. Perennials of wide distribution. Plants for the border are mostly hybrids and cultivars, but some of the smaller species are choice plants for the rock garden.

For the border: *A. caerulea* (the state flower of Colorado; North America; blue), *A. canadensis* (Canada; red and yellow), *A. crysantha* (North America; yellow), *A. longissima* (North America; yellow), *A. oxysepala* (Asia; violet and yellow), *A. vulgaris* (Common columbine; Europe; varying colors), etc., and hybrids and cultivars 'Clematis Flowered Hybrids,' 'McKana Giant Hybrids,' 'Mrs. Scott Elliot's Strain,' etc. See seedsmen's catalogs.

For the rock garden: *A. alpina* (Europe; blue), *A. bertolonii* (Europe; deep blue), *A. discolor* (Spain; blue and white), *A. pyrenaica* (Spain; blue), *A. viridiflora* (Siberia; brown and green), etc.

ARABIS   (Rock cress; Wall cress)

FAMILY: *Cruciferae*

The Greek word for Arabia, the natural home of these plants. All perennials for the rock garden.

A. albida (white) with varieties *coccinea* (crimson) and *flore-pleno*, *A. alpina* (white) with varieties *grandiflora* and *rosea* (rose), *A. aubrietioides* (purple lilac), *A. bryoides* (white), etc.

ARCTOSTAPHYLOS   (Bear berry)

FAMILY: *Ericaceae*

From the Greek *arktos*, bear, and *staphyle*, a bunch of grapes. Deciduous or evergreen shrubs for rock or wild garden.

A. alpina (deciduous; Asia, Europe, North America; pink), *A. californica* (evergreen; North America; pink), *A. tomentosa* (evergreen; California; white), *A. uva-ursi* (evergreen; Northern Europe, North America; pink).

ARCTOTIS

FAMILY: *Compositae*

From the Greek *arktos*, bear, and *ous*, an ear, referring to the shaggy seed pods of the annual form. One perennial and several annuals from South Africa; best treated as half hardy in most areas. Plants for the front border.

Annuals: *A. breviscapa* (orange), *A. grandis* (white, blue reverse), *A.* x *hybrida* (various colors) and cultivar 'Crane Hill' (blue and white), *A. laevis* (brownish red).

Perennial: *A. acaulis* (orange to carmine). See also *Venidio-Arctotis*.

ARENARIA

FAMILY: *Caryophyllaceae*

From the Latin *arena*, sand; referring to the fact that they are usually found growing in sandy soils.

Mostly perennials. Small, easy plants for the rock garden.

A. balearica (Balearic Islands, Mediterranean; white), *A. gracilis* (Europe; white), *A. ledebouriana* (Eastern Mediterranean areas; white), *A. montana* (Spain; white), *A. purpurascens* (Pyrenees; soft purple), etc.

ARGEMONE

FAMILY: *Papaveraceae*

From the Greek *argemos*, a white spot (cataract) on the eye, which this plant was once supposed to cure. A small genus of perennials from North America; usually grown as annuals.

A. grandiflora (white) and variety *lutea* (yellow), *A. mexicana* (Devil's fig; Prickly poppy; yellow), etc.

ARISTOLOCHIA   (Birthwort)

FAMILY: *Aristolochiaceae*

From the Greek *aristos*, best, and *lochia*, childbirth; referring to the one-time medical value of these plants. An infusion from them was also supposed to be a specific against the plague, and Gerard says "Dioscorides writeth that a dram weight of long birthwort drunk with wine and so applied is good against serpents and deadly things." Climbing plants chiefly interesting for their curious pipe-shaped flowers. Most are for the warm greenhouse but a few species may be grown outdoors in milder areas.

A. heterophylla (Europe; yellow), *A. macrophylla* (Dutchman's pipe, Pipe vine; North America; brownish yellow), *A. tomentosa* (North America; purple).

ARMERIA   (Sea pink; Thrift)

FAMILY: *Plumbaginaceae*

The old Latin name for pink, although this genus is in no way related to the true pink, a species of dianthus; and there appears still to be a certain confusion since some American writers classify *Armeria* with *Statice*, while in Europe *Statice* has been reclassified with *Limonium*. With these reservations it is given here as the gardener knows it; a genus of useful perennials for the front border and rock garden.

For the border: *A. maritima* (Britain; pink) and variety *laucheana* (bright red), *A. mauritanica* (Spain; rich carmine), *A. pseudoarmeria* (Portugal; rose) and cultivar 'Bee's Ruby,' etc.

For the rock garden: *A. caespitosa* (Spain; white to pink), *A. corsica* (Corsica; brick red), *A. splendens* (Spain; pale pink), etc.

ARNEBIA

FAMILY: *Boraginaceae*

From the Arabic name for this plant, *Arneb*. An annual and a perennial for the border introduced from Turkestan and Armenia.

Annual: *A. cornuta* (yellow, purplish spots).

Perennial: *A. echioides* (Prophet flower; yellow, brownish spots).

ARTEMISIA

FAMILY: *Compositae*

Named after the Greek Artemis, daughter of Zeus and sister to Apollo; the virgin huntress and goddess of wild life, childbirth and all young things. The plants themselves are a large genus   55 distributed all over the world; annuals, perennials and small shrubs. Nearly all have finely cut, aromatic foliage. The annuals are rarely seen but the perennials and sub-shrubs are useful for the border and rock garden. A selection only.

For the border, shrubs: *A. abrotanum* (Old man, Southernwood, Lad's love; Europe; grey), *A. absinthium* (Wormwood; Europe; yellow flowers), *A. arborescens* (Southern Europe; silvery foliage), *A. dracunculus* (Tarragon; Southern Europe; whitish green foliage), *A. lactiflora* (China; white), *A. ludoviciana* (North America; silvery, yellow), etc.

For the rock garden: *A. canescens* (Europe; silvery), *A. filifolia* (Mediterranean regions; silvery), *A. glacialis* (Alps; silvery, yellow), *A. lanata* (Southern Europe; silvery, yellow), etc.

ARUM

FAMILY: *Araceae*

Said to be the burning taste of these plants which gave them their name from the Arabic word *ar*, fire. The hardy species are handsome for the shaded wild garden. Not to be confused with the florists' Arum or Calla lily which is *Zantedeschia.*

A. italicum (Southern Europe; creamy white) and variety *pictum* (marbled leaves), *A. maculatum* (Cuckoo pint, Lords and ladies; Britain; yellowish green).

ASCLEPIAS   (Milkweed)

FAMILY: *Asclepiadaceae*

After Aesculapius, a Greek who was an authority on the medicinal properties of plants. Perennials;

mostly for the warm greenhouse, but some species may be grown in the open if they are given protection in severe winters.

A. incarnata (North America; silvery pink), *A. tuberosa* (North America; orange).

ASTER

FAMILY: *Compositae*

From the Greek *aster,* a star; describing the flowers. A large genus of well known perennials for the border and rock garden. A selection only.

For the border: *A. acris* (Southern Europe; lilac-purple), *A. amellus* (Italy; purple), *A. cordifolius* (North America; mauve) and variety *versicolor* (pink), *A. novae-angliae, A. novi-belgii* (Michaelmas daisy; North America) both with innumerable cultivars in a wide range of colors, *A. thomsonii* (Himalayas; pale blue), etc. See nurserymen's lists.

For the rock garden: *A. alpinus* (Europe; purple) and cultivar 'Beechwood,' *A. farreri* (China; violet blue), *A. natalensis* (Natal; blue), *A. yunnanensis* (China; lilac blue), etc. See nurserymen's lists.

For Annual or China Aster see *Callistephus.*

ASTILBE

FAMILY: *Saxifragaceae*

Thought to be derived from the Greek *a,* no, *stilbe,* brightness, since some of the earliest known species had almost colorless flowers. Perennials for the border and rock garden, of which the many modern cultivars are generally the most handsome. They are often known as *Spiraea.*

For the border: *A. davidii* (China; rose pink), *A.*   57

x *arendsii* hybrids including 'Amethyst,' 'Cattleya,' 'Fanal,' 'Feuer,' 'Red Sentinel,' etc. (Shades of lilac pink to crimson). See nurserymen's lists.

For the rock garden: *A. chinensis,* variety *pumila* (China; rose-lilac), *A. crispa* (probably a hybrid; salmon pink), *A. simplicifolia* (Japan; white to pink).

AUBRIETA   (Purple rock cress)

FAMILY: *Cruciferae*

Named after the French botanical artist, Claude Aubriet. Perhaps one of the most popular of all rock garden perennials. Several species but only one likely to be seen in cultivation and even that has been superseded by its numerous cultivars. Often incorrectly spelt 'aubretia.'

A. deltoidea (Southern Europe and Asia Minor; purple) with varieties *argenteo-variegata* (white variegated foliage) *and aureo-variegata* (yellow variegated foliage) and cultivars, 'Barker's Double,' 'Bonfire,' 'Dr. Mules,' 'Mrs. Rodewald,' etc. (Shades of pink to deep purple and crimson). See nurserymen's lists.

AURICULA

FAMILY: *Primulaceae*

From the Latin *auricula,* an ear; a reference to the shape of the leaves which resemble an animal's ear. So called 'Alpine auriculas' are probably derived from *Primula pubescens* and what are known as 'Florist's auriculas' from *Primula auricula.* Auricula itself is one of the 30 or so classes into which botanists now divide the genus PRIMULA, which see.

AZALEA is now classified with RHODODEN-DRON, and as this is a genus of generally large shrubs it is not included.

BAPTISIA   (False indigo)

FAMILY: *Leguminosae*

From the Greek *bapto,* to dye; as at one time some species were used for dyeing. Border perennials all from North America.

B. australis (blue), *B. leucantha* (white), *B. perfoliata* (yellow), *B. tinctoria* (yellow).

BARTONIA   See MENTZELIA

BEGONIA

FAMILY: *Begoniaceae*

Commemorating Michel Bégon (1638–1710), a Governor of Canada and patron of botany. Natives of moist tropical countries, most are greenhouse perennials, although many are becoming increasingly popular for summer bedding. The species have been so widely hybridised that naming them has become extremely complicated, but the following are probably among the parents of most modern begonias.

Tuberous rooted: *B. boliviensis* (Bolivia; scarlet), *B. clarkei* (Peru; rose), *B. davisii* (Peru; red), *B. rosaeflora* (Peru; rose), *B. veitchii* (Peru; red), etc.

Fibrous rooted: *B. albo-coccinea* (India; white and red), *B. acutifolia* (Jamaica; white), *B. evansiana* (China; pink), *B. incarnata* (Mexico; rose), *B. semperflorens* (Brazil; rose or white), etc.

Ornamental leaved: *B. albo-picta* (Brazil; leaves   59

green, spotted silver), *B. imperialis* (Mexico; deep green and light green), *B. metallica* (Brazil; green with a metallic lustre), *B. sanguinea* (Brazil; green and blood red), etc.

See nurserymen's lists.

BELLIS (Daisy; a corruption of)
(the Elizabethan name,)
(Day's eye)

FAMILY: *Compositae*

From the Latin *bellis*, pretty. Small old fashioned perennials.

B. perennis (Europe; white), *B. rotundifolia* and variety *caerulescens* (Algeria; tinged blue), *B. sylvestris* (Mediterranean area; pink), and cultivars (shades of pink).

BERGENIA

FAMILY: *Saxifragaceae*

Named for Karl August von Bergen, 1704–60, a German botanist. Hardy evergreen perennials at one time called *Megasea* and at another included with *Saxifraga*. Their large fleshy leaves are best suited for the wild garden.

B. cordifolia (Siberia; pink) and variety *purpurea* (reddish purple), *B. crassifolia* (Mongolia; pink), *B. delavayi* (China; purplish rose), etc., and cultivars 'Evening Glow,' 'Silberlicht,' etc.

BIGNONIA (Trumpet flower)

FAMILY: *Bignoniaceae*

Named after the Abbé Jean Paul Bignon, librarian to King Louis XIV. A climbing plant from the south eastern United States best treated as half hardy in most areas. One species only.

B. capreolata (syns. *Anisostichus capreolatus, Doxantha capreolata;* yellow flushed with red).

BOCCONIA   See under MACLEAYA.

BORAGO   (Borage)

FAMILY: *Boraginaceae*

From the Latin *burra,* rough; referring to the hairs on the stems and leaves. A small genus of plants extremely attractive to bees; *Borago officinalis* is used for flavoring drinks.

Perennial: *B. laxiflora* (Corsica; blue).

Annual: *B. officinalis* (Britain; bright blue).

BRACHYCOME   (Swan River daisy)

FAMILY: *Compositae*

From the Greek *brachys,* short, and *comus,* hair. Small annuals and perennials from Australia of which only one is usually cultivated; best treated as a half hardy annual in most areas.

B. iberidifolia and named hybrids, 'Little Blue Star,' 'Purple Splendour,' 'Red Star,' etc.

BRODIAEA

FAMILY: *Liliaceae*

Named after James Brodie, a Scottish botanist. A genus of hardy cormous plants from western North America. They grow abundantly in the meadows and woodlands of California and Oregon, where they are said to be known locally by such names as 'Fool's onion' and 'Nigger's toes.' It is a confused genus and some American botanists have split off a number of the species into separate genera; but there seems to be little general agreement and those listed here, with their synonyms, are as they are most usually seen in bulbsmen's catalogs.

B. californica (syn. *Hookera californica;* blue purple), *B. coronaria* (syn. *Hookera coronaria;* blue purple and white), *B. ida-maia* (Californian fire cracker; syns. *B. coccinea, Brevoortia ida-maia, Dichelostemma ida-maia;* scarlet and yellow), *B. laxa* (syn. *Tritelia laxa;* deep violet), *B. pulchella* (syns. *Dichelostemma pulchellum, Hookera pulchella;* violet blue) and variety *alba* (white).

B. uniflora is now known as *Iphion uniflorum;* see IPHEION.

BRUNNERA

FAMILY: *Boraginaceae*

Named for Samuel Brunner, a nineteenth century Swiss botanist. A small genus of perennials from Siberia and the Caucasus, of which the only species in cultivation is usually but incorrectly known as *Anchusa myosotidiflora.*

B. macrophylla (blue).

BUPHTHALMUM (Yellow ox-eye)

FAMILY: *Compositae*

From the Greek *bous,* an ox, and *ophthalmos,* an eye. Hardy herbaceous perennials from Southern Europe. Two species are in cultivation, best suited to the wild garden; both are yellow. *B. speciosum*

tiny plants only a few inches high to those of four and five feet. There are some annual forms, but most of the species in cultivation are biennials and perennials with hybrids and cultivars. All are characterised by the flower being more or less bell like in shape. The following is a much abbreviated selection of the species available.

For the border and bedding; perennials: *C. grandis* (Europe; blue), *C. lactiflora* (Caucasus; white to pale blue, sometimes pink) and many cultivars, *C. latifolia* (Britain; blue), *C. persicifolia* (Europe; white to lavender) and cultivars, etc.

Biennials: *C. medium* (Canterbury bell; Southern Europe; white, rose and purple) with varieties *calycanthema* (Cup and saucer) and *flore pleno* and seedsmen's strains, *C. patula* (Europe; violet), etc.

For the rock garden: *C. carpatica* (Eastern Europe; blue), *C. cochlearifolia* (syn. *C. pusilla;* European Alps; blue or white), and cultivars, *C. garganica* (Italy; blue), *C. lasiocarpa* (North America and Japan; blue), *C. pilosa* (Asia; blue), etc., and hybrids and cultivars.

For the wild garden: *C. barbata* (Europe; blue), *C. glomerata* (Europe; violet blue), *C. thyrsoides* (European Alps; yellow), etc.

See nurserymen's catalogs.

CANNA   (Indian shot)

FAMILY: *Cannaceae*

From *cana,* a cane or reed, referring to the tall, stiff flower scape. Handsome perennials from South America, the West Indies and Asia. Strictly greenhouse plants, but the one well known species (the hardiest) and its showy cultivars are much used for summer bedding and spot planting.

C. indica (West Indies; yellow and red) and named cultivars 'City of Portland,' 'Midas,' 'The President,' 'Wyoming,' etc., in shades of yellow, orange, rose and fiery scarlet.

CARDIOCRINUM

FAMILY: *Liliaceae*

From the Greek *kardia*, heart, and *krinon*, lily; of the heart-shaped leaves and the flowers respectively. Tall, handsome monocarpic plants (flowering and producing seed heads once only) from the Far East. Formerly included in the genus *Lilium*, and still often seen listed as *Lilium giganteum*.

C. cathayanum (China; greenish white), *C. cordatum* (Japan; creamy white), *C. giganteum* (Himalayas; white).

CARNATION   See DIANTHUS.

CATANANCHE   (Cupid's dart)

FAMILY: *Compositae*

From the Greek *katananke*, an incentive; said to refer to this plant's one-time use in love potions. The tradition may still persist in its common name. A small genus from Southern Europe of which only one perennial species is likely to be seen in cultivation.

C. caerulea (blue) with varieties *bicolor* (blue and white) and *alba* (white).

CELOSIA

FAMILY: *Amaranthaceae*

From the Greek *kelos*, burnt; suggesting the flame-like appearance of the flowers of some species. A genus of annuals from China generally treated as pot plants for the greenhouse, but often planted out for bedding in favorable localities.

C. argentea (white) with variety *linearis, C. cristata* (Cockscomb; red or yellow) and cultivars 'Empress,' 'Jewel Box,' 'Kurume Scarlet,' etc., *C. plumosa* (Prince of Wales' feathers; red or yellow), and cultivars 'Golden Plume,' 'Scarlet Plume,' etc. Some botanists consider that both *cristata* and *plumosa* are merely varieties of *C. argentea.*

CENTAUREA

FAMILY: *Compositae*

Named after the Centaurs of Greek legend; the classical name of a plant which Chiron, who was learned in medicine, is said to have used to heal a wound in his foot. A widely distributed genus of annuals and perennials.

Perennials: *C. dealbata* (Caucasus; rose) and variety *steenbergii* (crimson), *C. macrocephala* (Caucasus; yellow), *C. montana* (Pyrenees; blue), *C. pulcherrima* (Siberia; rose), etc.

Annuals: *C. americana* (North America; rose or purple), *C. cyanus* (Bachelor's button, Cornflower; dark blue) and cultivars in white and rose, *C. moschata* (Sweet Sultan; Asia; purple) and varieties *flava* (yellow) and *rosea* (pink), etc.

CENTRANTHUS

FAMILY: *Valerianaceae*

From the Greek *kentron*, a spur, and *anthos*, flower; the flower having a spur-like base. A small genus of annuals and perennials from the Mediterranean regions. The name is sometimes spelt Kentranthus.

Perennial: *C. ruber* (Red valerian; red) and varieties *albus* (white) and *atro-coccineus* (dark red).

Annual: *C. macrosiphon* (rose pink).

CEPHALARIA   (Giant scabious)

FAMILY: *Dipsaceae*

From the Greek *kephale*, a head; describing the flowers collected into a head. Large hardy perennials only suitable for the wild garden.

C. alpina (Europe; yellow), *C. tatarica* (Siberia; yellow).

CERASTIUM

FAMILY: *Caryophyllaceae*

From the Greek *keras*, a horn; referring to the horn-shaped seed capsules. Low growing carpeting plants; nearly all are invasive and easily become weeds.

C. alpinum (Europe; white) and variety *lanatum* (silvery foliage), *C. biebersteinii* (Asia Minor; white), *C. tomentosum* (Snow in summer; Europe; white).

CERATOSTIGMA

FAMILY: *Plumbaginaceae*

From the Greek *keras*, a horn, and *stigma*; in reference to the shape of the stigma. Perennials and small shrubs related to *Plumbago*. Those listed here are the hardiest; both are from China.

Perennial: *C. plumbaginoides* (syn. *Plumbago larpentae*; deep blue).

Shrub: *C. willmottianum* (sky blue).

CHEIRANTHUS

FAMILY: *Cruciferae*

The origin of the name is obscure but it may be from the Arabic *kheri*, meaning a fragrant red flower, together with the Greek *anthos*, flower. Perennials from Europe and North Africa although the popular bedding wallflowers, cultivars of *C. cheiri*, are grown as biennials. The genus is closely related to *Erysimum*.

C. allionii (probably a hybrid, or *Erysimum asperum* to some botanists; Siberian wallflower; Siberia; orange), *C. alpinus* (Scandinavia; yellow), *C. cheiri* (Wallflower, Gillyflower; Europe; variable colors) and many cultivars, 'Orange Bedder,' 'Blood Red,' 'Fire King,' 'Rose Queen,' etc., *C. x kewensis* (hybrid; orange and purple), *C. semperflorens* (syn. *C. mutabilis*; Morocco; purple). See nurserymen's catalogs.

CHELONE

FAMILY: *Scrophulariaceae*

From the Greek *kelone*, a tortoise; referring to the shell shape of the upper part of the flower. Perennials from North America, once included in the genus *Penstemon*.

C. barbata (syn. *Penstemon barbatus*; pink), *C. glabra* (white), *C. lyonii* (purple), *C. obliqua* (Shell flower; purple).

71

CHIASTOPHYLLUM

FAMILY: *Crassulaceae*

From the Greek *kiastos*, opposite, and *phyllon*, leaf; referring to the leaves placed opposite to each other on the stems. An attractive plant for the rock garden from the Caucasus; one species only but several names for it.

C. oppositifolium (syns. *Cotyledon oppositifolia, C. simplicifolia and Umbilicus oppositifolius*; evergreen, yellow).

CHIONODOXA   (Glory of the snow)

FAMILY: *Liliaceae*

From the Greek *chion*, snow, and *doxa*, glory; this genus flowering early as the snow melts on its native mountains. Small but attractive bulbous plants from the Mediterranean regions. Closely related to Scilla.

C. cretica (Crete; blue), *C. luciliae* (Turkey; blue, with pink and white forms), *C. sardensis* (Turkey; bright blue with rare pink and white forms).

CHRYSANTHEMUM

FAMILY: *Compositae*

From the Greek *chrysos*, gold, and *anthemon*, flower. A complex genus of over 100 species of annuals, perennials and sub shrubs from Africa, Asia, America and Europe, some of which are known as PYRETHRUM, which see. The well-known greenhouse and early flowering outdoor chrysanthemums are descendants of *C. indicum* and *C. sinense* from China and Japan, but the remaining species give a wide variety of subjects for bedding and the border, of which the following is a selection.

Perennials: *C. coccineum* (syn. *Pyrethrum roseum*, the pyrethrum of gardens; the Caucasus; variable pink to red), *C. leucanthemum* (Ox eye daisy; Europe, North America; white), *C. maximum* (Shasta daisy; Pyrenees; white), *C. sibiricum* (syns. *C. coreanum* and *Leucanthemum sibiricum*, from which the Korean chrysanthemums have been developed; Korea; variable colours), *C. uliginosum* (Eastern Europe; white), etc.

Annuals: *C. carinatum* (Painted daisy; Morocco; white and yellow) and a wide range of cultivars in zoned colors, *C. coronarium* (Southern Europe; yellow and white) and many cultivars, *C. segetum* (Corn marigold; Europe, Africa, Asia; golden yellow) and cultivars, etc. See seedsmen's and nurserymen's catalogs.

CIMICIFUGA

FAMILY: *Ranunculaceae*

From the Latin *cimex*, a bug, and *fugio*, to run away, since *Cimicifuga foetida* has insecticidal properties and was at one time used in Russia to drive away bed bugs. A small but widely distributed genus of tall hardy perennials for the shaded border. All are creamy or greenish white to white.

C. americana (North America), *C. elata* (China), *C. foetida* (Russia and Japan), *C. racemosa* (Eastern North America), etc., and cultivar 'White Pearl.'   73

CINERARIA

FAMILY: *Compositae*

From the Latin *cinereus*, ash colored; describing the appearance of the undersides of the leaves. All but a few species (which are not in cultivation) have now been transferred to the genus *Senecio*. The popular house and greenhouse plants, florists' cinerarias, are derived from *Senecio cruentus*, a perennial from the Canary Islands.

CISTUS   (Rock rose)

FAMILY: *Cistaceae*

From the Greek *kistos*, a rock rose. A genus of shrubs from the Mediterranean regions, where they may be seen blooming profusely in the mountains. Only the smallest are listed here, and none of these is likely to survive for long in hard winter areas.

C. crispus (rosy red), *C. loretii* (white with crimson blotches), *C. villosus* (magenta) and variety *alba* (white), with many hybrids and named cultivars. See specialist catalogs.

CLADANTHUS

FAMILY: *Compositae*

From the Greek *klados*, branch, and *anthos*, flower; referring to the flowers carried at the end of the branched stems. A genus of only one species; an annual from Spain and Morocco with strongly aromatic foliage. Usually treated as half hardy.

C. arabicus (syns. *C. proliferus*, *Anthemis arabica*; bright yellow).

CLARKIA

Family; *Onagraceae*

Commemorating Captain William Clark who, with Captain Meriwether Lewis (see also *Lewisia*), made an epic journey across America and through the Rocky Mountains early in the nineteenth century. Annuals from North America, of which the many brilliant cultivars have almost superseded the species.

C. elegans (rosy purple) and cultivars 'Brilliant,' 'Illumination,' 'Orange King,' 'Purple King,' etc., *C. pulchella* (white to shades of lilac).

CLAYTONIA

Family: *Portulacaceae*

Commemorating John Clayton, an eighteenth century American plant collector. Small spring flowering perennials for the moist rock garden.

C. australasica (Australia and New Zealand; white), *C. caroliniana* (North America; pink), *C. virginica* (North America; white)

CLEMATIS

Family: *Ranunculaceae*

From the Greek *klema,* a vine branch, referring to the climbing habit of most of the typical species. A large, diverse and widely distributed genus of which most of the well known garden forms are hybrids and cultivars. Many of the species, however, are equally rewarding and are becoming increasingly popular, and there are several little known semi-herbaceous species which given some support may be grown in the border.

75

Climbing: *C. alpina* (Europe; blue), *C. campaniflora* (Portugal; blue), *C. chrysocoma* (China; pink), *C. montana* (Himalayas; white to pink), *C. orientalis* (Persia; yellow to orange), *C. rehderiana* (China; pale yellow), *C. tangutica* (Siberia to China; yellow), and others.

Herbaceous and semi herbaceous: *C. heracleifolia* (China; blue) and variety *Davidiana* (pale blue), *C. integrifolia* (Europe; violet), *C. recta* (Europe; white), *C. texensis* (syn. *C. coccinea;* Texas; bright red), etc.

Most of the large flowered garden varieties are hybrids and cultivars, chiefly of *C. lanuginosa* (China), *C. macropetala,* (China), *C. viticella* (Spain), and others.

See specialist publications and nurserymen's lists.

CLEOME   (Spider flower)

FAMILY: *Capparidiceae*

From *kleome,* the Greek name for this plant. Large annuals from the West Indies, usually treated as half hardy but much hardier than is generally supposed.

C. hirta (pinkish mauve), *C. spinosa* (white to rose pink) with cultivars, 'Mauve Queen,' 'Pink Queen,' 'Rose Queen,' etc.

COBAEA   (Cup and saucer vine; Mexican ivy)

FAMILY: *Polemoniaceae*

Commemorating Father Barnadez Cobo, a Spanish Jesuit and naturalist who worked in Mexico during the eighteenth century. A small genus of climbing perennials from Mexico and

Central America. The one species usually seen in cultivation is treated as a half hardy annual in less favorable climates.

C. scandens (blue) with variety *aureo-marginata.*

CODONOPSIS

Family: *Campanulaceae*

From the Greek *kodon,* a bell, and *opsis,* like; describing the flowers. Attractive trailing perennials for the sheltered rock garden.

C. clematidea (Asia; blue and white), *C. convolvulacea* (China to Burma; lavender), *C. ovata* (Himalayas; blue), *C. vinciflora* (Tibet; lilac), etc.

COLCHICUM   (Autumn crocus; Meadow saffron)

Family: *Liliaceae*

The name is taken from Colchis, a province at the eastern end of the Black Sea; where the Argonauts found the Golden Fleece and Jason seduced the witch princess, Medea. Perhaps not surprisingly with this background the plant is beautiful but poisonous; natural source of the drug colchicene. Bulbous perennials, but not to be confused with the true crocus.

C. atro-purpureum (Europe; purple), *C. autumnale* (Europe; rosy purple) with varieties *alba* (white) and *roseum plenum* (rosy lilac), *C. byzantinum* (Asia Minor; pink and purple), *C. giganteum* (Asia Minor; pink and white), *C. speciosum* (Caucasus; lilac purple), etc.

COLEUS is a genus of house and greenhouse plants grown mainly for their colored foliage.

77

COLLINSIA

FAMILY: *Scrophulariaceae*

Named after Zaccheus Collins, a late eighteenth century American naturalist. Attractive annuals from North America, related to *Penstemon*.

C. bicolor (lilac and white) with varieties *alba* (white), *multicolor* (rose, lilac and white) and cultivar 'Salmon Beauty,' *C. grandiflora* (purple and blue), *C. verna* (white and blue).

COLLOMIA

FAMILY: *Polemoniaceae*

From the Greek *kolla*, glue; referring to a sticky secretion round the seeds. Brilliant annuals particularly attractive to bees.

C. biflora (syn. *C. coccinea;* Chile; scarlet), *C. grandiflora* (California; salmon).

CONVALLARIA   (Lily of the valley)

FAMILY: *Liliaceae*

From the Latin *convallis*, a valley. One species of tuberous rooted fragrant perennials widely distributed throughout Europe. They can become invasive.

C. majalis (white) with varieties *fortunei, rosea* (pink), and cultivar 'Fortin's Giant.'

CONVOLVULUS

FAMILY: *Convolvulaceae*

From the Latin *convolvo*, to entwine; as some of the species—more particularly those now included in other genera—grow in that way. Trailing annuals and perennials from Southern Europe and the

Mediterranean areas, best suited for the sunny rock garden. The perennials will not withstand frost and may themselves be best treated as annuals.

Perennials: *C. althaeoides* (rose pink), *C. cantabrica* (pale rose), *C. incanus* (pale blue), *C. mauritanicus* (deep blue), *C. tenuissimus* (pink).

Annual: *C. tricolor* (syn. *C. minor;* blue, pink and white) and cultivars 'Cambridge Blue,' 'Crimson Monarch,' 'Royal Ensign,' 'Royal Marine,' etc.

For *C. major* (Morning Glory) see *Ipomoea purpurea.*

COREOPSIS   (Tickseed)

FAMILY: *Compositae*

From the Greek *koris,* a bug or tick, and *opsis,* like; in reference to the appearance of the seed. Widely distributed throughout the United States; annuals and perennials, useful for the sunny border. The annual forms are sometimes listed as Calliopsis.

Perennials: *C. grandiflora* (yellow) and variety *flore-pleno,* cultivars 'Baden Gold,' 'Mayfield Giant,' 'Sunburst,' etc., *C. lanceolata* (yellow), *C. palmata* (orange), *C. pubescens* (syn. *C. auriculata;* yellow and crimson), *C. rosea* (pink), etc.

Annuals: *C. coronata* (orange and crimson), *C. drummondii* (yellow and crimson), and cultivars including 'Golden Crown,' *C. tinctoria* (yellow and crimson) and cultivars including 'Fire King,' 'Dazzler,' 'Star of Fire,' 'Garnet,' etc.

See seedsmen's lists.

CORYDALIS   (Fumitory)

FAMILY: *Fumariaceae*

From the Greek *korydalis,* the crested lark; a fanciful reference to the shape of the flowers. Small annuals and perennials, widely distributed throughout the Northern Hemisphere; useful for the rock garden.

Perennials: *C. allenii* (North America; pink and white), *C. cashmeriana* (Kashmir; blue), *C. cheilanthifolia* (China; yellow), *C. lutea* (Europe; yellow), *C. nobilis* (Siberia; yellow), *C. thalictrifolia* (China; yellow), etc.

Annual: *C. sempervirens*—a somewhat contradictory description for an annual (syn. *C. glauca;* Canada; pink to purple).

COSMOS

FAMILY: *Compositae*

From the Greek *kosmos,* beautiful. Annuals and perennials with feathery foliage; mostly from Mexico, and hence all usually treated as half hardy annuals in colder winter areas. Sometimes listed in catalogs under *Cosmea.*

C. atro-sanguineus (dark red), *C. bipinnatus* (rose to purple) and cultivars 'Early Express,' 'Sensation,' 'Sensation Radiance,' etc., *C. sulphureus* (yellow) and cultivars 'Early Orange Flare,' 'Yellow Klondyke,' 'Orange Ruffles,' etc.

CROCOSMIA

FAMILY: *Iridaceae*

Apparently from the Greek *krokos,* saffron, said to be a reference to the smell of the dried flowers when immersed in water; in fact the derivation is obscure. A small genus of cormous plants from South Africa, closely related to *Tritonia* and containing the hybrid race of garden montbretias which are derived from *C. aurea* x *C. pottsii*. They are best grown in a warm border and may need lifting and storing in hard winter areas. Montbretias generally are hardier than the species.

C. aurea (golden orange or orange red), *C. masonorum* (reddish orange), *C. pottsii* (yellow tinged red). Cultivars of Montbretia include 'Citronella,' 'His Majesty,' 'Earlham Hybrids,' etc. (shades of lemon to orange scarlet).

CROCUS

FAMILY: *Iridaceae*

From the Greek *krokos*, saffron; in reference to the true saffron crocus, *C. sativus*. Small cormous plants from the Mediterranean regions and Western Asia. The large Dutch hybrids and cultivars are too numerous and familiar to need description, but what does not appear to be so well known is that by a careful selection of the species, crocuses may be planted to flower from the autumn, through the winter in suitable areas, and on into the spring. The following is a selection only.

Autumn flowering (not to be confused with *Colchicum*): *C. asturicus* (Spain; purple), *C. byzantinus* (Eastern Europe; mauve to purple), *C. medius* (Sourthern France; lilac to purplish mauve), *C. ochroleucus* (the Lebanon; creamy white), *C. pulchellus* (Asia Minor; lavender), *C. sativus* (Saffron crocus; Southern Europe; variable rosy purple), etc.

Winter and early spring flowering: *C. aureus* (Greece to Asia Minor; golden yellow), *C. chrysanthus* (Greece to Asia Minor; yellow), *C. laevigatus* (Greece; lilac mauve), *C. tomasinianus* (Southern Europe; variable mauve), etc.

Spring flowering: *C. biflorus* (Italy to Asia Minor; white feathered purple), *C. candidus* (Asia Minor; white or pale yellow to orange), *C. corsicus* (Corsica; pale lilac feathered purple), *C. susianus* (South-west Russia; gold), *C. vernus* (European Alps and Pyrenees; variable white to deep purple), *C. versicolor* (Southern France; lilac feathered purple), etc.

Many of these species and others have numerous named cultivars; see bulbsmen's lists.

CYANANTHUS

FAMILY: *Campanulaceae*

From the Greek *kyanos*, dark blue, and *anthos*, a flower. Perennials for the rock garden from Central Asia; all are shades of blue.

C. incanus, *C. integer* (syn. *C. microphyllus*), *C. lobatus*, etc.

CYCLAMEN   (Sowbread)

FAMILY: *Primulaceae*

From the Greek *kyklos*, a circle; referring to the coiling of the flower stems in some species after flowering, the plant's own method of bringing the seed capsules down to soil level. A genus of dwarf tuberous plants from the Mediterranean regions. The valuable house plant is *C. persicum* and its cultivars, but there are many smaller species which in favorable localities will flower in the open in autumn and spring. The following is a selection of the hardier species.

Autumn flowering: *C. cilicium* (Asia Minor; pale rose pink), *C. europaeum* (Northern Italy; variable carmine), *C. neapolitanum* (syn. *C. hederaefolium;* Southern Italy, Greece; white to deep pink).

Spring flowering: *C. orbiculatum* (some botanists regard this as a group which includes *C. atkinsii* and others, while its syns. are *C. ibericum* and *C. vernum;* Southern Europe, Asia Minor; pink), *C. repandum* (Southern Europe; white to pink and crimson).

CYNOGLOSSUM   (Chinese forget-me-not)
                            (Hound's tongue)

FAMILY: *Boraginaceae*

From the Greek *kyon,* a dog, and *glossa,* tongue; alluding to the shape of the leaves. Biennials and perennials from China and the Himalayas, best treated as biennials for the border and rock garden.

For the border: *C. amabile* (pink, blue, or white), *C. nervosum* (blue) and cultivars 'Blue Bird,' 'Firmament,' 'Pink Firmament,' etc.

For the rock garden: *C. wallichii* (sky blue).

DAHLIA

FAMILY: *Compositae*

Commemorating Andreas Dahl, a Swedish botanist who was a pupil of Linnaeus. Tuberous rooted perennials from Mexico, first brought into cultivation about the end of the eighteenth century;   *83* usually treated as half hardy. Few if any of the original species may be seen today and all of the vast range of magnificent modern specimens are hybrids and cultivars.

D. coccinea (scarlet) the parent of the modern single dahlia, *D. coronata* (scarlet), *D. gracilis* (orange red), *D. juarezii* (scarlet) parent of the cactus types, *D. merckii* (lilac and yellow), *D. variabilis* (variable); the last two between them are parents of most double and decorative dahlias, and the latter of the fancy and pompom classes.

DAPHNE

FAMILY: *Thymelaeaceae*

Named after Daphne, in Greek mythology a river nymph who was pursued by Apollo and, after praying for help, was transformed into a laurel bush—which then became sacred to Apollo. A widely distributed genus of small deciduous and evergreen shrubs. The following is a selection.

D. alpina (deciduous; European Alps; white), *D. Cneorum* (Garland flower; evergreen; Europe; pink), *D. Mezereum* (deciduous; Europe and Siberia; purplish red or white), *D. pontica* (evergreen; Asia Minor; yellowish green), *D. retusa* (evergreen; China; white touched with rose), etc. There are many varieties and hybrids; see nurserymen's lists.

DARLINGTONIA   (Californian pitcher plant)

FAMILY: *Sarraceniaceae*

Commemorating Dr. William Darlington, an American botanist. An insectivorous perennial from California, said to be quite hardy except in severe winter areas.

D. *californica* (yellow, green and brown).

DATURA is omitted since the best species are shrubs for the greenhouse.

DELPHINIUM

FAMILY: *Ranunculaceae*

From the Greek *delphin*, a dolphin; referring rather fancifully to the flower buds having some resemblance to a dolphin's head. A large genus of annuals, biennials and perennials of wide distribution. As a result of the work of the great plant breeders—Reinelt, Blackmore and Langdon, Bishop, Legro, Steichen, and many others—their hybrids and cultivars are among the most stately of all garden plants. The whole subject is now extremely complex, but the tall growing giant cultivars are derived chiefly from D. *elatum*, the elegant strain known as "Connecticut Yankees" from the *belladonna* types, and the strains of annual delphiniums known as larkspur from D. *ajacis* and D. *consolida*. Many of the species, however, are unduly neglected.

Perennials: D. *cardinale* (California; bright red), D. *elatum* (North America and Europe to Siberia; blue) and many hybrids and cultivars, D. *formosum* (Asia Minor; purple blue; probably also contributed to garden hybrids), D. *grandiflorum* (syn. D. *chinense*; North America to Siberia; violet blue) and cultivars 'Azure Fairy,' 'Blue Butterfly,' etc. D. *nudicaule* (California; yellow to red), D. *welbyi*　85

(Abyssinia; pale blue), *D. zalil* (Afghanistan and Persia; yellow) etc. and strains including 'Blackmore and Langdon's,' 'Connecticut Yankees,' 'Pacific Giants,' etc. (white through gentian blue to intense purple), and in the near future further developments in shades of yellow, orange and red.

Annuals (Larkspur): *D. ajacis* (Europe; variable, white through blue and rose pink to violet), *D. consolida* (Europe; violet) and strains 'Blue Spire,' 'Coral King,' 'Giant Imperial,' etc.

See specialist publications and nurserymen's catalogs.

DENTARIA

FAMILY: *Cruciferae*

From the Latin *dens*, a tooth; referring to the toothlike scales on the roots. Uncommon and useful little perennials for the shade.

D. bulbifera (syn. *Cardamine bulbifera*; Britain; pale mauve), *D. digitata* (Southern Europe; purple rose), *D. laciniata* (North America; white), *D. polyphylla* (syn. *D. enneaphylla*; European Alps; creamy white).

DIANTHUS

FAMILY: *Caryophyllaceae*

From the Greek *dios*, divine, and *anthos*, a flower; divine flower, or flower of Zeus. A large genus of annual, biennial and perennial plants which falls into three main groups; pinks, carnations and dianthus proper. Again the modern hybrids and cultivars are now extremely complex but the carnation is ultimately descended from *D. caryophyllus* and the pink probably from *D. plumarius*, which was introduced to Britain by the monks from Normandy about the year 1100 when it was much in demand for its clove fragrance used for flavoring wine; hence one of its many common names, "Sops in wine," "Pink," "Carnation" and at one time "Gillyflower," are names which are used for both of these and other species. It is a genus with long historical traditions, but the present day gardener's main interest will probably lie in the many cultivars for the border and bedding and the smaller species which are invaluable for the rock garden. The following is a much shortened selection.

Perennials for the border: *D. carthusianorum* (Europe; crimson), *D. caryophyllus* (Carnation, Pink; Europe; variable), *D. knappii* (Europe; yellow), *D. plumarius* (Pink; Europe; variable), etc.

Perennials for the rock garden: *D. alpinus* (European Alps; rose to crimson), *D. deltoides* (Britain to Japan, white to crimson), *D. microlepis* (Bulgaria; pink or white), *D. neglectus* (Pyrenees; rose), *D. sternbergii* (Southern Europe; rose), etc.

Biennial, or usually treated as such, for bedding and the border: *D. barbatus* (Sweet William; Southern Europe; variable).

Annual, or usually treated as such, for bedding and the border: *D. chinensis* (Chinese or Indian pinks, annual carnations with *D. caryophyllus*; Portugal to China; variable).

There are many varieties, hybrids and cultivars of these; see nurserymen's catalogs and specialist publications.

DICENTRA

FAMILY: *Fumariaceae*

From the Greek *di*, two, and *kentron*, a spur; referring to the curious shape of the flowers. A genus of decorative perennials, formerly known as *Dielytra*, for the border and rock garden.

For the border: *D. eximia* (North Carolina; purple) with variety *alba* (white), *D. formosa* (North America; red) and cultivar 'Bountiful,' *D. spectabilis* (Dutchman's breeches; Bleeding heart, Lyre flower; Siberia and Japan; rose red) and variety *alba.*

For the rock garden: *D. cucullaria* (North America; white and yellow), *D. oregana* (Oregon; creamy pink), *D. peregrina* (Japan; white to rose pink).

DICTAMNUS   (Dittany;)
            (Burning bush)

FAMILY: *Rutaceae*

The ancient Greek name, *dictamnos*; the plant once growing on Mount Dicte, Greece, where according to legend Zeus was born. One species; a curious perennial, since the plant gives off a volatile oil from the upper parts of its stems which may be ignited on hot days and will burn without harming the plant.

D. albus (syn. *D. fraxinella*; white) with varieties *caucasicus* (syn. *giganteus*; purple) and *rubra* (rosy red).

DIERAMA   (Wand flower)

FAMILY: *Iridaceae*

From the Greek *dierama*, a funnel; describing the
88   shape of the individual flowers hanging from long,

slender stems. Perennials from South Africa, not quite hardy in wet cold winters.

D. pendulum (lilac), *D. pulcherrimum* (syn. *Sparaxis pulcherrima;* rose purple), variety *album* (white) and cultivars 'Heron,' 'Skylark,' etc.

DIGITALIS   (Foxglove)

FAMILY: *Scrophulariaceae*

From the Latin *digitus,* a finger; individual flowers resembling the finger of a glove. Perennial and biennial plants widely distributed in Europe; the natural source of the drug digitalin.

Perennials: *D. davisiana* (bronze), *D. laevigata* (bronze yellow), *D. lutea* (yellow), etc.

Biennials or treated as such: *D. ferruginea* (rust red), *D. grandiflora* (syn. *D. ambigua;* yellow blotched brown), *D. lanata* (cream), *D. purpurea* (common foxglove; purple). There are hybrids and cultivars from these species which include the 'Shirley Strains,' 'Excelsior hybrids,' 'Foxy,' etc.

DIMORPHOTHECA   (Namaqualand daisy;)
(Star of the veldt)

FAMILY: *Compositae*

From the Greek *di,* two, *morphe,* shape, and *theca,* seed; because the flower produces two different shapes of seed. Annuals and perennials from South Africa, all usually treated as half hardy annuals in less favorable climates; most of them have the annoying habit of only opening their flowers in bright sunshine. Considerable confusion in naming exists and according to some authorities the species should be split between the genera

Castalis, Chrysanthemoides, Dimorphotheca and Osteo-spermum. The following appear to be the more usual descriptions but are given with some reservation.

D. aurantiaca (syns. D. calendulacea, D. sinuata; Cape marigold; brilliant orange) and hybrids 'Goliath,' 'Lemon Queen,' 'White Beauty,' etc., D. barberiae (syn. Osteospermum barberiae; rosy lilac), D. pluvialis (white, purple reverse) and variety ringens (violet ring around the central disk), etc.

DODECATHEON   (American cowslip)

FAMILY: Primulaceae

From the Greek dodeka, twleve, and theos, a god; an ancient name signifying Flower of the Twelve Gods. Attractive perennials for the rock garden from North America, related to Primula and Sol-danella.

D. clevelandii (California; violet), D. hendersonii (Oregon; purple and yellow), D. jeffreyi (California; purple rose), D. latifolium (North-west America; bright pink), D. meadia (Shooting star; Eastern North America; rosy purple and white), etc.

DORONICUM   (Leopard's bane)

FAMILY: Compositae

From the Arabic name for this plant, doronigi. Perennials from Europe and Asia. The sap from the roots of some species is said to be poisonous. All are yellow.

D. austriacum (Austria), D. caucasicum (Caucasus), and variety magnificum, D. clussi (Europe), D. cor-datum (Asia Minor), D. plantagineum (Europe with variety 'Harper Crewe,' etc., and the double hybrid 'Frühlingspracht.'

DOROTHEANTHUS

FAMILY: *Aizoaceae*

Named in honor of Frau Dorothea Schwantes, wife of a German botanist. Succulent plants from South Africa often found listed with the larger and more diverse genus *Mesembryanthemum.* Dwarf trailing subjects strictly for the greenhouse, but treated as half hardy annuals they are becoming increasingly popular for fast and brilliant ground cover in the summer months.

D. bellidiformis (white and pink to orange), *D. criniflorum* (syn. *Mesembryanthemum criniflorum;* variable), *D. gramineus* (carmine), *D. tricolor* (white and purple).

DOUGLASIA   A small genus of rock plants mainly from North America; now rarely seen in cultivation.

DRABA   A genus of over 250 annuals, biennials and perennials from the colder areas of the world. Those cultivated are best grown in the alpine house.

DRACOCEPHALUM   (Dragon's head)

FAMILY: *Labiatae*

From the Greek *drakon,* a dragon, and *kephale,* the head; a reference to the gaping flower mouth. Annuals and perennials for the front border from Europe and Asia. Not to be confused with "False dragon head," which is *Physostegia.*

Perennials; *D. austriacum* (Europe; blue), *D. grandiflorum* (Siberia; blue), *D. hemsleyanum* (Tibet; light blue), *D. sibiricum* (syn. *Nepeta macrantha*; Siberia; blue), *D. speciosum* (Himalayas; lilac), etc.

Annual: *D. moldavica* (Moldavian balm; blue).

DRYAS   (Mountain avens)

FAMILY: *Rosaceae*

From the Greek *dryas*, a dryad or wood nymph. A small genus of attractive evergreen perennials from the mountain regions of the Northern Hemisphere; for the sunny rock garden.

D. drummondii (North America; yellow), *D. octopetala* (Northern Europe, North America; white), and a hybrid, *D* x *suendermannii* (D. drummondii x D. octopetala; yellow opening white).

ECCREMOCARPUS   (Chilean glory flower)

FAMILY: *Bignoniaceae*

From the Greek *ekkremes*, pendant, and *karpos*, fruit; describing the hanging seed vessels. One species only in cultivation; an attractive climbing plant best treated as a half hardy annual in all but the mildest areas.

E. scaber (Chile; scarlet and yellow) with varieties *aureus* (golden yellow), and *rubra* (red).

ECHINACEA   (Cone flower)

FAMILY: *Compositae*

From the Greek *echinos*, a hedgehog; referring to the prickly bracts behind the flower head. Two species of tall border perennials from North America. See also *Rudbeckia*.

E. angustifolia (syn. *Rudbeckia angustifolia;* purplish red), *E. purpurea* (syn. *Rudbeckia purpurea;* purple), and cultivars 'The King,' 'Robert Bloom,' 'White Lustre,' etc.

ECHINOPS   (Globe thistle)

FAMILY: *Compositae*

From the Greek *echinos,* a hedgehog, and *opsis,* like; describing the spines which surround the flower heads. Hardy perennials sometimes used as spot plants in the border or wild garden.

E. humilis (Asia; blue), with variety *nivalis* (white) and cultivar 'Taplow Blue,' *E. ritro* (Southern Europe; steel blue), *E. sphaerocephalus* (Europe and Western Asia; silver).

ECHIUM   (Viper's bugloss)

FAMILY: *Boraginaceae*

From the Greek *echis,* a viper; referring either to the one time belief that this plant was a remedy for the bite of a viper, or to the supposed resemblance of its seeds to a viper's head. Annual, biennial and perennial plants from the Mediterranean areas and the Canary Islands; *E. vulgare* is also found in Britain.

Perennials: *E. albicans* (rose becoming violet).

Biennials: *E. plantagineum* (bluish purple), *E. vulgare* (purple or blue).

Annual: *E. creticum* (violet).

EDRAIANTHUS

FAMILY: *Campanulaceae*

From the Greek *edraios,* stemless, and *anthos,* flower. Attractive perennials for the sunny rock garden, from the Mediterranean regions; closely related to *Wahlenbergia*.

E. dalmaticus (purple-blue), *E. dinaricus* (syn. *Wahlenbergia dinarica*; violet), *E. graminifolius* (purple), *E. serpyllifolius* (syn. *Wahlenbergia serpyllifolia*; purple-violet), *E. tenuifolius* (violet blue), etc.

ENDYMION   (Common bluebell)

FAMILY: *Liliaceae*

Named after Endymion, the beautiful youth of Greek mythology loved by Selene, the moon; by her contrivance he was thrown into a perpetual sleep on Mount Latmos and she descended every night to embrace him. Nevertheless a somewhat restless species; bulbous plants generally included in *Scilla*, often named as *Hyacinthus*, commonly known as bluebells, of the English woods, but botanically a small separate genus. They come from Britain, Spain and Portugal.

E. hispanicus (Spanish bluebell; blue to white or pink), *E. non-scriptus* (English bluebell, distinct from the bluebell of Scotland which is *Campanula rotundifolia*; blue, sometimes pink or white) and hybrids 'Blue Queen,' 'Rose Queen,' etc.

EPILOBIUM   (Rosebay willow herb;)
             (Fireweed)

FAMILY: *Onagraceae*

From the Greek *epi*, upon, and *lobos*, a pod; from the flowers appearing to grow on the seed pod. Almost any of the species can become dangerous weeds and are best avoided but one may be tried with caution on the sunny rock garden.

94   *E. obcordatum* (California; rosy purple).

EPIMEDIUM

FAMILY: *Berberidaceae*

From *epimedion,* the ancient Greek name of obscure meaning used by Pliny. A widely distributed genus of attractive hardy perennials for the large rock garden or wild garden.

E. alpinum (Bishop's hat; Southern Europe; rose-purple and yellow), *E. diphyllum* (syn. *Aceranthus diphyllus;* Japan; white), *E. grandiflorum* (Japan; yellow or rose to violet), *E. pinnatum* and variety *colchicum* (Barrenwort; Caucasus and Persia; yellow), etc.

ERANTHIS   (Winter aconite)

FAMILY: *Ranunculaceae*

From the Greek *er,* spring, and *anthos,* flower. Small perennials which are best naturalised beneath trees. Their distinctive ruffs of glistening green leaves and bright flowers are among the first to appear in early spring. All are yellow.

E. cilicia (Greece and Asia Minor), *E. hyemalis* (Western Europe), hybrids *E* x *tubergenii* and 'Guinea Gold,' etc.

EREMURUS   (Foxtail lily)

FAMILY: *Liliaceae*

From the Greek *eremos,* solitary, and *oura,* a tail; referring to the appearance of the single flower spike. Imposing but sometimes difficult plants, mostly from Western Asia and the Himalayan regions; they are best suited to large, sheltered gardens.

E. bungei (Persia; yellow), *E. himalaicus* (Himalayas; white), *E. robustus* (Turkestan; pink), *E. spectabilis* (Siberia; yellow to orange), etc., with hybrids   95

'Highdown Hybrids,' 'Shelford Hybrids' and cultivars 'Dawn,' 'Flair,' 'Sir Arthur Hazelrigg,' etc.

ERICA   is a genus best suited to mass effects in
the shrub garden and is not included.

ERIGERON   (Fleabane)

FAMILY: *Compositae*

The derivation is obscure but possibly from the
Greek *eri*, early, and *geron*, old; referring to the
greyish leaves of some species, or the white-haired
seed head. A widely distributed genus of hardy
perennials, useful for the rock garden or front
border.

For the border: *E. aurantiacus* (Turkestan; or-
ange), *E. coulteri* (North America; white or mauve),
E. macranthus (North America; violet), *E. philadel-
phicus* (North America; lilac pink), *E. speciosus*
(North America; violet blue), etc., and cultivars
'Bressingham Strain,' 'Darkest of All,' 'Foerster's
Liebling,' 'Gartenmeister Walther,' etc.

For the rock garden: *E. alpinus* (European Alps;
purple), *E. aureus* (North America; golden yellow),
E. mucronatus (Mexico; white, pale and deep pink
all together), *E. uniflorus* (North America; white or
purple), etc.

ERINUS

FAMILY: *Scrophulariaceae*

From the Greek *eri*, early, since the plants flower
in the spring. One hardy perennial species for the
rock garden from the Pyrenees.

E. alpinus (rosy purple) with varieties *albus*

(white), *carmineus* (carmine), and cultivars 'Dr. Haenaele,' 'Mrs. C. Boyle,' etc.

ERITRICHIUM is a genus of beautiful but extremely difficult rock plants for the connoisseur's alpine house.

ERODIUM   (Heron's bill)

FAMILY: *Geraniaceae*

From the Greek *erodios,* a heron, since the style and ovaries resemble the head of a heron. Hardy perennials, closely related to the true *Geranium,* or crane's bill. For the front border and rock garden.

For the border: *E. absinthoides* (Europe, Asia Minor; white), *E. manescavii* (Pyrenees; red), *E. pelargoniflorum* (Turkey; white marked purple), etc.

For the rock garden: *E. chrysanthum* (Greece; yellow), *E. corsicum* (Corsica; red), etc.

ERYNGIUM   (Sea holly)

FAMILY: *Umbelliferae*

From the ancient Greek name *eryngeon,* but its meaning now is obscure. A widely distributed genus of over 200 species. Some make handsome spot or accent plants and are also useful for floral decoration; the following are among the best and hardiest. All are spiny and shades of more or less metallic blue.

E. alpinum (Central Europe), *E. amethystinum* (Southern Europe), *E. bourgatii* (Pyrenees), *E. giganteum* (Caucasus), *E. maritimum* (the true Sea holly; Britain).   97

ERYSIMUM   (Alpine wallflower)

FAMILY: *Cruciferae*

From the Greek *erus*, to draw up, since some species are said to produce blisters. A genus of annual, biennial and perennial plants resembling *Cheiranthus*, to which they are closely related.

Perennials: *E. dubium* (syn. *E. ochroleucum*; Europe; pale yellow), *E. rupestre* (Asia Minor; sulphur yellow), etc.

Biennials: *E. allionii* (syn. *Cheiranthus allionii*), *E. arkansanum* (Arkansas and Texas; yellow), *E. asperum* (see *Cheiranthus allionii*; North America; orange), *E. linifolium* (syn. *Cheiranthus linifolius*; Spain; rosy lilac), etc.

Annual: *E. perofskianum* (Afghanistan; reddish orange).

ERYTHRONIUM

FAMILY: *Liliaceae*

From the Greek *erythros*, red, the flower color of the European species. Graceful bulbous perennials mainly from North America although the common species, *E. dens-canis* (Dog's tooth violet) is a native of Europe. Of great value for the lightly shaded rock garden.

E. americanum (golden yellow), *E. californicum* (creamy white), *E. citrinum* (white and lemon yellow), *E. dens-canis* (Dog's tooth violet; variable white to pinkish mauve), *E. hendersonii* (lilac), *E. revolutum* (Trout lily; white to deep pink), *E. tuolumnense* (yellow), etc., and cultivars from these, 'Franz Hals,' 'Purple King,' 'White Beauty,' 'Pagoda,' etc.

ESCHSCHOLTZIA

FAMILY: *Papaveraceae*

Commemorating Johann Friedrich von Eschscholtz, a doctor and naturalist with a Russian expedition to North-west America in the early nineteenth century. A small genus of Western American annuals from two species of which the many modern strains and named varieties are derived.

E. caespitosa (yellow), *E. californica* (California poppy; yellow to orange) with variety *alba flore pleno* (double white) and many cultivars, 'Monarch Art Shades,' 'Carmine King,' 'Golden Glory,' etc. See seedsmen's lists.

EUPATORIUM

FAMILY: *Compositae*

Named after Mithridates Eupator, a king of Pontus about 115 B.C. who is said to have discovered an antidote to poison in one of the species; when he was taken by his enemies he preferred death to captivity, but he had fortified himself against poison so strongly that he could not poison himself and had to order a slave to stab him. The medicinal value of some of these plants has been known throughout the ages and among their many local common names "Joe-pye weed" is said to have come from an Indian herb doctor. A genus of over 400 species widely distributed throughout America and Europe. Of those in cultivation the best are greenhouse shrubs and not many of the hardy perennials have any particular value, but a few may be grown in the wild garden.   *99*

E. cannabinum (Hemp agrimony; Europe; purple lilac) and variety *plenum* (double pink), *E. purpureum* (Joe-pye weed; North America; pale purple), *E. rugosum* (syns. *E. ageratoides, E. urticaefolium;* White snakeroot; North America; white), etc.

EUPHORBIA

FAMILY: *Euphorbiaceae*

Named after Euphorbus, physician to a king of ancient Mauretania. A genus of about a thousand species, distributed mainly in the temperate regions and showing a wide diversity of form; it includes annuals, biennials, perennials, shrubs, trees and succulents. All exude an irritating milky sap when the stems are broken; a few are poisonous if eaten. Nevertheless some are striking garden plants by reason of their showy bracts arranged around the small and insignificant flowers. The best known example is probably the handsome house plant, Poinsettia; *E. pulcherrima.* The following is a selection of the hardier species for the border.

Perennials: *E. griffithii* (Himalayas; orange), *E. polychroma* (Europe; yellow), *E. sikkimensis* (Eastern Himalayas; yellow), *E. wulfenii* (Europe; yellow), etc.

Biennial: *E. lathyris* (Caper spurge; Europe; green), etc.

Annuals: *E. heterophylla* (Mexican fire plant; North and South America; red), *E. marginata* (North America; white), etc.

FELICIA

FAMILY: *Compositae*

From the Latin *felix,* cheerful; a reference to the bright flowers. A few species of small annuals and perennials from South Africa. The perennials are greenhouse plants and the annuals listed here are best treated as half hardy in most areas.

F. bergeriana (Kingfisher daisy; bright blue), *F. tenella* (syn. *F. fragilis;* violet blue).

FILIPENDULA   (Meadow sweet)

FAMILY: *Rosaceae*

From the Latin *filum,* a thread, and *pendulus,* hanging; a reference to the root tubers of *F. hexapetala* which hang on thread-like roots. Widely distributed perennials growing in damp places. There is still some confusion between this genus and *Spiraea* to which it formerly belonged. The larger species, listed here, are best suited for moist positions in the border or wild garden.

F. camtschatica (syns. *Spiraea gigantea* and *S. camtschatica;* Kamtchatka to Manchuria; shades of pink), *F. hexapetala* (syns. *Spiraea filipendula* and *Ulmaria filipendula;* Europe; white), *F. purpurea* (syn. *Spiraea palmata;* Japan, carmine), *F. rubra* (syns. *Spiraea lobata, S. rubra, S. venusta* and *Ulmaria rubra;* Queen of the prairie; North America; peach pink), etc. All of these species have varieties.

FRITILLARIA

FAMILY: *Liliaceae*

From the Latin *fritillus,* a dice box, referring to the checkered markings on the flowers of some species. A large and widely distributed genus of bulbous plants, mostly of curious colors and patterning. The "Crown imperial" is an old fashioned flower of cottage gardens, but few of the other species are widely grown. The following is a selection of those most likely to be offered by bulbsmen. *F. recurva* is included since it is said to be the glory of the genus; when it can be obtained— and, even more rarely, when it can be grown.

F. acmopetala (Asia Minor; jade green and brownish maroon), *F. bithynica* (syn. *F. citrina;* Asia Minor; citron yellow and green), *F. imperialis* (Crown imperial; Europe; orange, varying to yellow and red in its varieties and cultivars), *F. meleagris* (Chequered lily, Guinea-hen flower, Snake's head lily; Europe; variable from white to brownish maroon and chequered) with cultivars 'Aphrodite,' 'Artemis,' 'Charon,' 'Saturnus,' etc., *F. persica* (Syria to North-west Persia; greenish and dull purple), *F. pyrenaica* (Pyrenees; variable but generally chequered brownish purple, sometimes yellow inside), *F. recurva* (California and Oregon; scarlet flecked with orange), etc.

FUCHSIA

FAMILY: *Onagraceae*

Commemorating Leonard Fuchs, a sixteenth century German botanist and professor of medicine. A genus of a hundred or so species and literally thousands of hybrids raised over the last century. Most are for the greenhouse and strictly all are shrubs; but the hardier species and their cultivars listed here are becoming increasingly grown as garden plants in areas where the winters are not too severe.

F. magellanica (South America; scarlet and purple) with varieties including *alba* (pale pink), *riccartonii* (scarlet and violet), *F. parviflora* (Mexico; crimson and coral red), *F. reflexa* (Mexico; cerise), etc., and cultivars which are usually hardier and smaller than the species, 'Caledonia,' 'Corallina,' 'Dunrobin Bedder,' 'Madame Cornelissen,' etc. See nurserymen's catalogs. For "Californian fuchsia" see *Zauschneria*.

FUNKIA   See Hosta

GAILLARDIA   (Blanket flower)

FAMILY: *Compositae*

Named after M. Gaillard de Marentonneau, a French patron of botany during the eighteenth century. A small genus of annuals and perennials, all from North America. The species are now mostly superseded by named cultivars, but a few may still be offered.

Perennial: *G. aristata* (syn. *G. grandiflora;* zoned red and yellow) and the many cultivars arising from this, 'Burgundy,' 'Copper Beauty,' 'Firebird,' 'Goblin,' etc.

Annual: *G. pulchella* (crimson and yellow) with varieties *lorenziana* and *picta,* cultivar 'Indian Chief' and strains 'Double Fireball,' 'Double Mixed,' etc.   *103*

GALANTHUS   (Snowdrop;)
                    (Fair maids of February)

FAMILY: *Amaryllidaceae*

From the Greek *gala*, milk, and *anthos*, flower; describing the flowers, all of which are white touched with green. Small, hardy bulbous plants from Britain, Europe and Asia. *G. nivalis*, the common snowdrop, is the most grown but some of the others are better.

G. byzantinus (South east Europe), *G. caucasicus* (Caucasus), *G. elwesii* (Greece and Turkey), *G. nivalis* (Britain and Europe), and varieties, etc.

GALAX   (Wand plant)

FAMILY: *Diapensiaceae*

From the Greek *gala*, milk. Probably referring to the color of the flowers. One species only from North America, useful for ground cover. The richly tinted autumn leaves are valued for floral arrangements.

G. aphylla (white).

GALTONIA   (Cape hyacinth; Spire)
                    (lily; Summer hyacinth)

FAMILY: *Liliaceae*

Named after Sir Francis Galton, the nineteenth century anthropologist and a cousin of Charles Darwin. A small genus of tall, hardy bulbous plants from South Africa of which only one species is usually seen in cultivation.

104   *G. candicans* (syn. *Hyacinthus candicans*; white).

GAURA

FAMILY: *Onagraceae*

From the Greek *gauros,* superb, referring to the beauty of the flowers. A genus of annual and perennial plants from North America, of which the only two species in cultivation are both usually grown as annuals.

Perennial: *G. lindheimeri,* (rosy white).

Annual: *G. parviflora* (yellow).

GAZANIA   (Treasure flower)

FAMILY: *Compositae*

Commemorating Theodore of Gaza, a fifteenth century translator of the botanical works of Theophrastus. Perennials from South Africa best treated as half hardy annuals in less favorable climates; the flowers only open in the sunshine, but when they do they have an enamelled, jewellike quality. *G. splendens* is probably the hardiest. The species hybridise freely and a wide range of colors with contrasting zonal markings is now offered.

G. longiscapa (golden yellow), *G. pavonia* (yellow and brown), *G. rigens* (orange), *G. splendens* (orange, black and white), named cultivars, and *G. hybrida* (variable colors).

GENTIANA

FAMILY: *Gentianaceae*

Named after Gentius, a king of Illyria who is said to have first used the plant medicinally. A large and widely distributed genus, mostly of   *105* hardy perennials. A few are handsome plants for the border but nearly all are for the rock garden, and among these are some of the most beautiful of all such plants. Of *Gentiana* Reginald Farrer—probably still one of the greatest authorities on alpine plants—says "Take it all in all Gentiana offers the rock garden more glory than any other race and more persistently denies it. To please Primula is possible, to cope with Campanula is even comfortable, but there is no jesting with a Gentian. . . ." The following is a selection of the more amenable species, although it does include one rare and difficult aristocrat from America; *Gentiana crinita* the "Fringed gentian," a biennial which is worth any amount of trouble to grow if seed or plants can be obtained.

For the border: *G. andrewsii* (Bottle gentian; Eastern North America; blue), *G. asclepiadea* (Willow gentian; Europe; blue or white), *G. lutea* (Europe; yellow).

For the rock garden: *G. acaulis* (according to Farrer a description which covers at least five other related species, and to modern botanists probably a hybrid, but still usually seen listed under this name; Alpine regions; variable to rich blue), *G. gentianella* (sometimes used as another name for *acaulis*, but considered by Farrer to be an old garden form of this group and certainly a hybrid itself; rich blue), *G. gracilipes* (China; blue), with variety *alba* (white) and several hybrids, *G. saxosa* (New Zealand; white), *G. septemfida* (Asia; blue), *G. sino-*
ornata (China; deep blue), *G. verna* (Europe includ-

ing Britain; clear blue), etc.

Biennial: *G. crinita* (Fringed gentian; North America; clear blue).

Gentiana is a specialist's genus and many do not agree on its nomenclature. See nurserymen's lists.

GERANIUM   (Crane's bill)

FAMILY: *Geraniaceae*

From the Greek *geranos,* a crane, because the seed pod is thought to resemble a crane's head and beak. Perennials for the border and rock garden from the temperate regions all over the world. Not to be confused with so-called geraniums for the greenhouse and bedding, which are in fact *Pelargonium.* A short selection only.

For the border: *G. armenum* (Asia Minor; magenta), *G. grandiflorum* (Himalayas; bluish mauve), *G. phaeum* (Mourning widow; Europe; dark purple), *G. pratense* (Britain; blue), *G. sanguineum* (Bloody crane's bill; Britain; red), etc., and varieties and cultivars.

For the rock garden: *G. argenteum* (European Alps; pink), *G. cinereum* (Pyrenees; pink), *G. endressii* (Pyrenees; rose), *G. sanguineum* variety *lancastriense* (Britain; pale pink), etc., and varieties of these. See nurserymen's lists.

GEUM   (Avens)

FAMILY: *Rosaceae*

Probably from the Greek *geno,* to taste or to give an agreeable flavor, since the roots of some species are aromatic. A genus of perennials for the border and rock garden from America, Europe and the *107*

Near East. Most of the border forms now grown are hybrids and cultivars.

For the border: *G.* x *borisii* (hybrid; orange), *G. bulgaricum* (Europe; yellow), *G. chiloense* (syn. *G. coccineum;* Chile; scarlet) and cultivars 'Lady Stratheden,' 'Princess Juliana,' 'Fire Opal,' 'Red Wing,' etc.

For the rock garden: *G. reptans* (Europe; yellow), *G. rivale* (Britain; rose), etc.

GILIA

FAMILY: *Polemoniaceae*

Commemorating Felipe Luis Gil, an eighteenth century Spanish botanist. Annuals, biennials and shrubby perennials of which the few species in cultivation are from California. Useful for the border and, the dwarfs *G. hybrida* and *G. micrantha,* for edging.

Perennials: *G. californica* (pink), *G. montana* (syn. *Linanthus montanus;* white).

Biennial: *G. rubra* (variable, yellow to scarlet).

Annuals: *G. achilleifolia* (purplish blue), *G. capitata* (lavender), *G. densiflora* (syn. *Leptosiphon densiflorus;* lilac), *G. hybrida* (syn. *Leptosiphon hybridus;* variable colors), *G. micrantha* (rose pink), *G. tricolor* (lavender and white, darker throats).

GLADIOLUS

FAMILY: *Iridaceae*

From the Latin *gladiolus,* a little sword; describing the shape of the leaves. Upwards of 150 species of cormous perennials from the Mediterranean area, and North, Tropical and South Africa. Few of them are in cultivation, however, having been superseded by the several thousand cultivars listed today. These are derived from less than a dozen original South African ancestors—*G. cardinalis, G. oppositiflorus, G. psittacinus, G. purpureo-auratus, G. papilio*, etc.,—and are divided into their own distinct classes; see publications of the various Gladiolus Societies and bulbsmen's catalogs. Nevertheless some of the species are of garden interest and while most of these are best in the cool greenhouse a few are hardy in milder winter areas.

G. atro-violaceus (Turkey; violet-purple), *G. byzantinus* (Eastern Mediterranean area; magenta crimson), *G. illyricus* (Mediterranean areas; magenta purple), *G. segetum* (Southern Europe; rosy purple).

GLOBULARIA

FAMILY: *Globulariaceae*

From the Latin *globulus*, a little globe; the shape of the flower heads. There is one shrub, but those listed here are dwarf sub-shrubs and perennials for the sunny rock garden. All come from the Mediterranean regions and all are shades of blue.

G. cordifolia, G. incandescens, G. nudicaulis, G. repens, G. trichosantha, etc.

GODETIA

FAMILY: *Onagraceae*

Commemorating Charles Godet, a nineteenth century Swiss botanist. Annuals related to the evening primrose, *Oenothera*, in which genus they were originally included. The original species have been largely superseded by their cultivars.

G. dasycarpa (South America; mauve) and some lavender flowered cultivars, *G. amoena* (California; rose and crimson), *G. grandiflora* (California; crimson and white) and cultivars from these 'Firelight,' 'Kelvedon Glory,' 'Rich Pink,' 'Rosy Queen,' 'Carmine Glow,' etc. (shades of salmon to carmine), *G. viminea* (California; purple) and cultivars 'Lavender,' 'Lavender Gem,' etc. See seedsmen's catalogs.

GRINDELIA

FAMILY: *Compositae*

Commemorating D. H. Grindel, an eighteenth century German botanist. Biennials, perennials and shrubs from North and South America; the few in cultivation are usually treated as biennials. All are yellow.

G. chiloensis (syn. *G. speciosa*; South America), *G. integrifolia* (North America), *G. robusta* (Californian gum plant; California).

GUNNERA

FAMILY: *Haloragidaceae* (or *Gunneraceae*)

Commemorating J. Ernst Gunnerus, an eighteenth century Norwegian bishop and botanist. Ornamental perennials mostly from the Southern Hemisphere and some of such massive proportions that usually they are only planted in public grounds for landscape effects; there is, or was, one specimen in England under which several people might stand and shelter from the rain. At the other extreme the same genus includes small, somewhat insignificant species which may be used for ground cover in moist areas.

Giant species: *G. chilensis* (syn. *G. scabra*; Chile; reddish brown inflorescence), *G. manicata* (Brazil; greenish inflorescence).

Dwarf species: *G. magellanica* (Southern Chile).

GYPSOPHILA

FAMILY: *Caryophyllaceae*

From the Greek *gypsos,* chalk, and *phileo,* to love, since the plants grow best on chalky soils. Useful annuals and perennials for the rock garden and border. They are mainly natives of the eastern Mediterranean areas, but some extend through Asia to Japan.

Perennials for the border: *G. pacifica* (syn. *G. oldhamiana;* Japan; pink), *G. paniculata* (Baby's breath; Europe; white), with varieties *compacta* and *flore-pleno,* and cultivars 'Bristol Fairy,' 'Rosy Veil,' etc.

Annual for the border: *G. elegans* (Caucasus; white), and varieties, etc.

Perennials for the rock garden: *G. cerastioides* (Himalayas; white touched with red), *G. dubia* (Eastern Europe; white flushed pink), *G. repens* (European Alps; white to rose), etc.

HELENIUM   (Sneezeweed)

FAMILY: *Compositae*

Said to be named after Helen of Troy, since according to legend these flowers sprang from the ground where her tears fell. All of the species from which the modern garden cultivars are derived, however, come from North America. They are useful late summer and autumn perennials for the border.

111

H. autumnale (yellow) and variety *pumilum, H. bigelovii* (yellow and brown), *H. hoopesii* (yellow) and cultivars 'Butterpat,' 'Chipperfield Orange,' 'Crimson Beauty,' 'Goldlackzwerg,' etc.

HELIANTHEMUM (Sun rose)

FAMILY: Cistaceae

From the Greek *helios,* the sun, and *anthemon,* a flower. The genus includes some annuals and perennials but most of the numerous hybrids and cultivars have arisen from a few species of subshrubs. They are of great value for the rock garden or front border.

H. alpestre (European Alps; yellow), *H. apenninum* (Europe and Asia Minor; white), *H. nummularium* (syn. *H. vulgare;* Europe; yellow), etc., and cultivars in a wide range of colors from white to yellow, orange and carmine; 'Beech Park Scarlet,' 'Ben Attow,' 'Ben Ledi,' 'Butterball,' 'Lemon Queen,' etc. See nurserymen's catalogs.

HELIANTHUS (Sunflower)

FAMILY: *Compositae*

From the Greek *helios,* the sun, and *anthos,* a flower. Coarse annuals and perennials from most parts of the temperate regions of the Northern Hemisphere. The giant sunflower, *H. annuus,* has economic importance, yielding bird seed, edible oil and a yellow dye, and *H. tuberosus* is the edible Jerusalem artichoke. All come from North America.

Perennials: *H. atro-rubens* (yellow, purplish red disk) and cultivar 'The Monarch,' *H. decapetalus* (sulphur yellow) and cultivars 'Capenoch Star,' 'Loddon Gold,' 'Soleil d'Or,' etc., *H. laetiflorus* (yellow) and cultivar 'Miss Mellish,' *H. tuberosus* (yellow), etc.

Annuals: *H. annuus* (Giant Sunflower; yellow), variety *flore-plenus* and cultivars 'Primrose,' 'Russian Giant,' 'Tall Chrysanthemum Flowered,' etc., *H. debilis* (yellow) and cultivars 'Autumn Beauty,' 'Dazzler,' 'Excelsior' (shades of yellow zoned with red), etc.

HELICHRYSUM   (Everlasting flower;)
                         (Immortelle)

FAMILY: *Compositae*

From the Greek *helios*, the sun, and *chrysos*, gold; referring to the yellow flowers of some species. A large and widely distributed genus of generally rather undistinguished annuals, perennials and shrubs. They can be used in the border and rock garden, and the flowers of some may be dried for winter floral arrangements.

Perennials or sub-shrubs for the border: *H. angustifolium* (Southern Europe; yellow), *H. lanatum* (South Africa; white foliage, yellow), *H. splendidum* (South Africa; greyish foliage, yellow).

Perennials for the rock garden: *H. bellidioides* (New Zealand; white), *H. frigidum* (Corsica; silvery foliage, white), *H. orientale* (South-east Europe; yellow), etc.

Annuals: *H. bracteatum* (Straw flower; Australia; yellow or pink) with varieties *alba* (white) and *monstrosum*; cultivars and strains of *H. b. monstrosum* include 'Fireball,' 'Golden,' 'Rose,' etc.

HELIOPSIS

FAMILY: *Compositae*

From the Greek *helios*, the sun, and *opsis*, like; describing the flowers. Rather coarse but useful perennials for the border from North America. All are shades of yellow to orange.

H. helianthoides (False sunflower), *H. scabra* (Orange sunflower), with varieties *incomparabilis* and *patula*, *H. zinniiflora*, etc., and cultivars 'Golden Plume,' 'Orange King,' 'Sonnenschild,' 'Summer Sun,' etc.

HELIOTROPIUM   (Cherry pie)

FAMILY: *Boraginaceae*

From the Greek *helios*, the sun, and *trope*, to turn. According to the legend Apollo loved Clytie but forsook her for her sister, Leucothoe; Clytie then pined away and on her death Apollo—presumably in a fit of remorse—changed her into a flower which is always turning towards the sun, heliotrope. A genus of tender shrubs from Peru one of which, *H. arborescens*, is usually treated as a half hardy annual for summer bedding.

H. amplexicaulis (syn. *H. anchusifolium;* lavender), *H. arborescens* (syn. *H. peruvianum;* heliotrope and white) and cultivars 'Lemoine's Giant,' 'Lord Roberts,' 'President Garfield,' 'White Lady,' etc.

HELIPTERUM

FAMILY: *Compositae*

From the Greek *helios*, the sun, and *pteron*, a wing or feather; in reference to the plumed seed pappus.

A small genus of annuals and shrubby perennials

from Australia. The annuals are sometimes offered by seedsmen and usually treated as half hardy. Their dried flowers are useful for winter decoration.

H. humboldtianum (yellow), *H. manglesii* (syn. *Rhodanthe manglesii;* pink and white), *H. roseum* (syn. *Acroclinium roseum;* shades of pink) with variety *grandiflorum* (variable), and cultivar 'Red Bonnie,' etc.

HELLEBORUS

FAMILY: *Ranunculaceae*

From the Greek *helein,* to kill, and *bora,* food; some of the species being poisonous. Hardy perennials, widely distributed throughout Southern Europe and Western Asia. They often retain their leathery foliage throughout the winter and their flowers are long lasting.

H. argutifolius (syn. *H. corsicus;* Mediterranean regions; apple-green flowers), *H. foetidus* (Britain; green tinged with purple), *H. niger* (Christmas rose; Europe; white), and cultivars (sometimes flushed with pink), *H. odorus* (Central Europe; greenish yellow), *H. orientalis* (Lenten rose; Greece; variable pink to blackish purple) and cultivars, *H. viridis* (Europe; green), etc. See nurserymen's catalogs.

HELONIAS   (Swamp pink; Stud flower)

FAMILY: *Liliaceae*

From the Greek *helos,* a marsh; the natural habitat of this plant. One species only. A tuberous rooted perennial from Eastern North America;   *115* useful for moist, heavy soil in the shaded wild garden.

H. bullata (purplish pink).

HELXINE   (Baby's tears;)
          (Mind-your-own-business)

FAMILY: *Urticaceae*

Possibly from the Latin *helix*, ivy, since the plant creeps. A single species from Corsica forming thick mats of tiny bright green leaves. In Victorian times it was grown as ground cover in large conservatories, or in pots in cottage windows—where it is safest. Once allowed to break loose in the garden it might well be called gardener's tears since it can quickly become a dangerous and ineradicable weed.

H. soleirolii.

HEMEROCALLIS   (Day lily; Lemon lily)

FAMILY: *Liliaceae*

From the Greek *hemero*, a day, and *kallos*, beauty, as each individual flower lasts only for one day, but since every head will often carry up to two dozen buds in fact the plant remains in bloom for weeks. A genus of perhaps the most adaptable of all perennials from Eastern Asia and Southern Europe, in which again the species have been almost superseded by the hundreds of brilliant cultivars raised in America, Europe and Japan—to which more are being added each year.

H. aurantiaca (Japan; orange yellow), *H. citrina* (South-eastern Europe; citron yellow), *H. fulva* (Europe and Japan; orange brown) and varieties,

H. middendorffii (Siberia and Japan; golden yellow), *H. minor* (an interesting dwarf species; Siberia and Japan; yellow), etc. Cultivars include 'Ballet Dancer,' 'Black Prince,' 'Hyperion,' 'Ophir,' 'Pink Prelude,' 'Red Torch,' etc., and some of the finest are derived from hybrids raised by Dr. A. B. Stout of the New York Botanical Garden. See specialist nurserymen's catalogs.

HEPATICA

FAMILY: *Ranunculaceae*

From the Greek *hepar*, liver, which the lobed leaves are supposed to resemble. A few species of perennials growing wild in woodland areas all over the north temperate zones. They were once included in the genus *Anemone*. Useful for the shaded rock garden.

H. americana (syns. *H. triloba, Anemone hepatica;* Europe, North America; deep lavender blue) with varieties *alba* (white) and *rubra flore-pleno* (double pink), *H. transsilvanica* (syns. *Anemone transsilvanica, A. angulosa;* Eastern Central Europe; variable lavender to pink), etc.

HERACLEUM   after Hercules; coarse plants which are best regarded as giant weeds.

HERMODACTYLUS   (Snake's head iris)

FAMILY: *Iridaceae*

From the name *Hermes* (Mercury) and *dactylos*, a finger; literally the "fingers of Hermes" in reference to the shape of the roots. Hermes is said to have invented the lyre; he was the god of luck and   *117* wealth, patron of merchants—and also of thieves. A single species of perennial plants which originated in the Mediterranean area but is now said to grow wild in some parts of Britain. It was once included in the genus *Iris* and known as *Iris tuberosa.*

H. tuberosus (flowers greenish yellow and dull violet).

HESPERIS

FAMILY: *Cruciferae*

From the Greek *hesperos*, evening, when the flowers of some species become sweetly scented. A genus of biennials and perennials from Southern Europe. They are similar in form to *Matthiola* and *Cheiranthus*, but taller and useful for the border.

Perennials: *H. matronalis* (Dame's rocket, Dame's violet, Sweet Rocket; variable white to lilac) with varieties *candidissima* (pure white) and *purpurea* (purple).

Biennial: *H. tristis* (variable white to purple), etc.

HESPEROYUCCA (Chaparral yucca;)
(Quijote plant)

FAMILY: *Liliaceae*

From the Greek *hesperos*, western, and *yucca*; the yucca, or the plant like a yucca, from the west. A small genus from the United States and Mexico sometimes included under *Yucca.* There is only one species known to be in cultivation.

H. whipplei (syn. *Yucca whipplei*; California to Arizona; flowers creamy white).

HEUCHERA

FAMILY: *Saxifragaceae*

Named in honor of J. H. Heucher, a German professor of medicine and a botanist in the seventeenth and eighteenth centuries. A genus of decorative, near evergreen perennials from North America. All are perennials and valuable for the front border, edging, and ground cover.

H. americana (Alum root; red) and hybrids of this with *H. sanguinea, H. micrantha* (pale yellow), *H. pubescens* (deep pink), *H. sanguinea* (Coral bells; pink) and varieties *alba* (white), *atro-sanguinea, grandiflora, rosea, splendens* (shades of rose to deep red), *H. villosa* (pink), etc. There are many cultivars which include 'Bressingham Hybrids,' 'Cascade,' 'Coral Cloud,' 'Flambeau,' 'Pearl Drops,' 'Pluie de Feu,' 'Queen of Hearts,' etc.

HEUCHERELLA

FAMILY: *Saxifragaceae*

A bigeneric hybrid between HEUCHERA and TIARELLA, which see.

HIBISCUS

FAMILY: *Malvaceae*

The ancient Greek name given to a mallow-like plant. A genus largely from the tropical regions and comprised mainly of large shrubs and greenhouse subjects, but including a few annual and perennial species which make large, handsome border plants in suitably mild areas.

Perennial: *H. moscheutos* (Rose mallow; Southern

United States; variable white and pink to red) and cultivars 'Southern Belle,' etc.

Annuals, or treated as such: *H. diversifolius* (Pacific Islands and Australia; white to yellow, maroon center), *H. trionum* (syn. *H. africanus;* Bladder ketmia; Africa and North America; white to yellow, purple center), etc.

HIERACIUM   (Hawkweed)

FAMILY: *Compositae*

From the ancient Greek *hierax,* a hawk; Pliny, the Roman naturalist, believed that hawks fed on this plant to strengthen their eyesight. Somewhat weedy perennials from Europe and North Africa which can become troublesome by seeding about too freely. Suitable for dry, poor positions.

H. aurantiacum (orange red), *H. maculatum* (yellow), *H. villosum* (silvery foliage; yellow), etc.

HIPPOCREPIS

FAMILY: *Leguminosae*

From the Greek *hippos,* a horse, and *crepis,* shoe; referring to the shape of the seed pod. A small genus from Europe, North Africa and Western Asia. Only one species is likely to be seen in cultivation, a trailing perennial for the rock garden which has the single advantage of being indestructibly easy.

H. comosa (Britain; yellow).

HOSTA   (Plantain lily)

FAMILY: *Liliaceae*

Commemorating Nicolaus and Joseph Host, 120 Austrian botanists of the late eighteenth and early nineteenth centuries. A genus of hardy perennials mainly valued for their large decorative leaves, also known as Funkia. They may be used for the front border, for the wild garden, or as spot plants in terrace and patio gardening. There are many species in cultivation and the following is a selection; unless otherwise stated all these come from Japan.

H. albo-marginata (syn. *H. lancifolia,* var. *albo-marginata;* leaves with white margins, flowers violet), *H. decorata* (narrow margin, dark lilac), *H. fortunei* (pale lilac), *H. plantaginea* (China; white), *H. tardiflora* (lilac), etc. See nurserymen's catalogs.

HOUSTONIA

FAMILY: *Rubiaceae*

Commemorating Dr. William Houston, an early eighteenth century English botanist. A small genus of hardy perennials from North America. Those in cultivation are excellent plants for the shady rock garden.

H. caerulea (Bluets; pale blue) with variety *alba* (white) and cultivar 'Millard's Variety' (deeper blue), *H. purpurea* (white to pink), *H. serpyllifolia* (white).

HUNNEMANNIA   (Bush eschscholtzia;)
                (Mexican tulip poppy;)
                (Santa Barbara poppy)

FAMILY: *Papaveraceae*

Named after John Hunneman, an English botanist of the eighteenth century. A genus of a single species, perennial in its native Mexico but usually   *121* treated as a half hardy annual for the border in less favorable climates.

H. fumariifolia (silvery foliage; yellow flowers).

HUTCHINSIA

FAMILY: *Cruciferae*

Named after Miss Ellen Hutchins, an Irish botanist of the late eighteenth and early nineteenth centuries. A small genus of annuals and perennials, of which there are two perennials in cultivation. They are good rock garden plants; both from the European Alps and both white.

H. alpina, H. brevicaulis.

HYACINTHUS

FAMILY: *Liliaceae*

From Hyakinthos who, in Greek mythology, was loved by Apollo and accidentally killed by the god when playing quoits, owing to the jealous intervention of Zephyrus; whereupon hyacinths sprang up from where the beautiful youth's blood fell. Some thirty species of bulbous plants, mainly from the Mediterranean regions, of which there are now very few in cultivation. The showy hybrids used for forcing and bedding have been raised mostly in Holland from *H. orientalis,* and as long ago as 1686 the Leyden Botanic Garden listed 35 varieties. The following species, however, are sometimes offered by bulbsmen.

H. amethystinus (Pyrenees; bright blue and a rare white form), *H. azureus* (Asia Minor; azure blue), *H. orientalis* (Eastern Mediterranean region; variable). For the numerous named florists' varieties see bulbsmen's catalogs.

HYMENOCALLIS

FAMILY: *Amaryllidaceae*

From the Greek *hymen*, a membrane, and *kallos*, beauty; referring to the cup-like membrane which unites the stamens. A genus of bulbous plants from America, mainly the south. They are usually regarded as warm greenhouse subjects although one at least is almost hardy; even with this, however, the bulbs should be lifted and stored over winter in all but the mildest areas.

H. calathina (syn. *H. narcissiflora*; Amerindian lily, Spider lily; the Andes; white).

HYPERICUM   (St. John's wort)

FAMILY: *Hypericaceae* or *Guttiferae*

The Greek name, of obscure meaning; it is possibly *hyper*, over, and *ereike*, heath, in reference to the natural habitat of many of the species. A large genus of annuals, perennials, sub-shrubs and shrubs widely distributed throughout the Northern hemisphere. Most of those in cultivation are more suitable for shrub planting, and the few listed here are some of the dwarf species which are useful for ground cover and good in the rock garden. All are bright yellow.

H. fragile (Greece), *H. olympicum* (Asia Minor), *H. repens* (Asia Minor), etc.

HYSSOPUS   (Hyssop)

FAMILY: *Labiatae*

The ancient Greek name for the plant. A semi-evergreen perennial from Southern Europe; sometimes used for flavoring and for centuries highly regarded as a cure for coughs and chest com-

123

plaints. It is also an attractive flowering plant for the border.

H. officinalis (blue, with red and white forms).

IBERIS (Candytuft)

FAMILY: *Cruciferae*

From the ancient name for Spain; Iberia. Annual and sub-shrubby perennials, these mostly evergreen, from Spain and the Mediterranean regions. Useful for the front border and rock garden.

Perennials: *I. saxatilis* (white tinged purple), *I. sempervirens* (white).

Annuals: *I. amara* (white), *I. umbellata* (purple) with varieties *albida* (white), *purpurea* (dark purple) and many hybrid strains 'Empress,' 'Dunnet's Crimson,' 'Giant Pink,' etc.

IMPATIENS

FAMILY: *Balsaminaceae*

From the Latin *impatiens,* in reference to the way in which the seed pods of some species burst and scatter their seeds when touched; *I. balsamina,* "Touch me not," being the best known. Some hundreds of species, annuals, biennials and perennials, mostly from the mountains of Asia and Africa. Few are in cultivation and these are mostly greenhouse perennials, but some of their modern cultivars are offered as half hardy annuals by seedsmen.

Perennials, strictly for the greenhouse but cultivars treated as half hardy annuals: *I. holstii* 124 (Africa; red), *I. sultanii* (Busy Lizzie; Zanzibar; scarlet) and hybrid strains 'Glow in Shadow,' 'Imp Series,' etc.

Annual, half hardy: *I. balsamina* (Balsam, Touch-me-not; Asia; white to rose and scarlet) and cultivars 'Improved Camellia Flowered,' etc.

Annual, hardy: *I. roylei* (Himalayas; purple), etc.

INCARVILLEA

FAMILY: *Bignoniaceae*

Commemorating Pierre d'Incarville, an eighteenth century French Jesuit missionary to China. A small genus of perennials, hardy in milder and drier winter areas. Already commemorating one French Jesuit *I. delavayi* also takes the name of another, also stationed in China; Père Jean Marie Delavay (1834–1895) who discovered the first plant.

I. delavayi (China; rose pink), *I. grandiflora* (China; deeper color blotched white) and variety *brevipes* (crimson), *I. olgae* (Turkestan; purple).

INULA

FAMILY: *Compositae*

Probably a corruption of *helenium; I. helenium,* "elecampane," being the *Inula campana* of medieval Latin. An undistinguished genus of perennials, mostly from Europe. All are shades of yellow to orange and best suited to the wild garden.

I. ensifolia (Southern Europe), *I. helenium* (Elecampane; Europe), *I. oculis-Christi* (Europe), *I. royleana* (Himalayas), etc.

IONOPSIDIUM   (Violet cress)

FAMILY: *Cruciferae*

From the Greek *ion*, a violet, *opsis*, appearance, and *idion*, small. Useful little annuals from Portugal, Sicily, and Algeria for odd spaces in the rock garden or between paving stones. A small genus, with only one species in cultivation.

I. acaule (white, tinged violet).

IPHEION   (Spring starflower)

FAMILY: *Liliaceae*

The origin of the name is obscure. A genus of small bulbous plants from North and South America once included in *Brodiaea*. There is only one species cultivated. This is useful for pockets in the rock garden or for naturalising, but in suitable conditions it multiplies so rapidly that it may become invasive.

I. uniflorum (white to pale blue).

IPOMOEA   (Morning Glory)

FAMILY: *Convolvulaceae*

From the Greek *ips*, bindweed, and *homoios*, like; referring to the twining habit of growth. A genus of several hundred species (including the sweet potato, *I. batatas*) mainly from the tropics, Africa, Asia and Australia. Most are greenhouse plants but Morning Glory is treated as a half hardy annual for outdoors; this and its varieties may be seen referred to either as *I. hederacea* or *I. purpurea* but is more probably a hybrid or cultivar of one or both of them. The similar plant, Moonflower, so called because it opens at night, is sometimes known as *I. mexicana*, but this appears to be a synonym of *Calonyction aculeatum*, also normally a greenhouse plant.

I. hederacea (Tropical America; blue to purple), *I. purpurea* (Tropical America; purple) and cultivars 'Improved Heavenly Blue,' 'Scarlett O'Hara,' 'Giant Pink,' 'Rose Marie,' etc.

See also QUAMOCLIT.

IRIS

FAMILY: *Iridaceae*

From the Greek *iris*, a rainbow; or Iris, messenger of the gods whose path was the rainbow. A vast genus distributed over almost all of the north temperate zone from Algeria to Siberia, and Japan to North America. The species are all perennial, falling into many different classes and subdivisions, and ranging from those only a few inches high for the choicest parts of the rock garden to others of four feet and more for the wild garden. The best known are the complex hybrids described as Tall Bearded Irises, and as a result of the work of iris specialists all over the world, but more particularly in America, these are among the most stately flowers we have. But many of the species are no less beautiful, and it would be perfectly possible to fill every aspect of a fairly large garden with them and have one or another in flower continuously from early December, in mild areas, to July of the following year. Any short selection is bound to be arbitrary and to leave out many which are equally desirable; and in describ- 127 ing them as merely for the border, the water-side or the rock garden no reference is made to their general classifications—bearded, beardless, Japanese, regelia, bulbous, etc. See specialist works and the publications of American, British and other iris societies.

For the border: *I. chrysographes* (South West China; red to violet purple), *I. forrestii* (Yunnan; yellow), *I. longipetala* (California; violet and white), *I. ochroleuca* (Asia Minor; white and yellow), *I. sibirica* (Central Europe and Russia; blue to purple), *I. unguicularis* (Algeria; winter flowering; mauve or white), *I. versicolor* (North America; blue to red purple), etc.

For the rock garden: *I. cristata* (South-eastern United States; mauve, white and orange), *I. gracilipes* (Japan; lilac), *I. histrioides* (Asia Minor; blue), *I. innominata* (California; variable golden yellow), *I. reticulata* (Asia; purple), *I. tectorum* (China; mauve), etc.

For the waterside: *I. clarkei* (Western China; blue-purple), *I. fulva* (Southern United States; coppery red), *I. kaempferi* (South Korea; purple, widely variable in cultivars), *I. laevigata* (South Korea; dark blue or white), etc.

See specialist catalogs.

IXIA   (African corn lily)

FAMILY: *Iridaceae*

From the Greek *ixia*, bird lime; referring to the sticky sap. Small cormous plants from South 128 Africa, hardy only in mild winter areas.

I. campanulata (syn. *I. speciosa;* purple and crimson), *I. maculata* (orange), *I. patens* (pink), *I. viridiflora* (bluish green).

IXIOLIRION
FAMILY: *Amaryllidaceae*

From the Greek *ixia*, bird lime, and *leiron*, a lily; literally a lily like an ixia. A small genus of bulbous plants from Asia allied to *Alstroemeria*. Useful in the border but like the ixias they are only hardy in the mildest areas.

I. montanum (syn. *I. pallasii;* blue) and variety *ledebouri* (paler blue).

JEFFERSONIA
FAMILY: *Berberidaceae*

Commemorating Thomas Jefferson (1743–1826), President of the United States. Perennials from North America and Asia, and choice plants for the shady rock garden.

J. diphylla (North America; white), *J. dubia* (Asia Minor; soft lilac blue).

KIRENGESHOMA
FAMILY: *Saxifragaceae*

A tall-growing Japanese plant which has retained its native name. There is only one species in cultivation.

K. palmata (yellow).

KITAIBELIA
FAMILY: *Malvaceae*

Named in honor of Professor Paul Kitaibel, a Hungarian botanist. A tall, handsome perennial   *129*

from the Lower Danube area. Only one species.
K. vitifolia (white to rose).

KNIPHOFIA   (Red hot poker;)
            (Torch lily)

FAMILY: *Liliaceae*

Commemorating an eighteenth century German professor of medicine, Johann Heironymus Kniphof. Perennials from Africa and Madagascar which are often known as Tritoma and are most aptly described by their common names. Showy plants for the border, but they may need some winter protection in colder districts.

K. caulescens (South Africa; buff to red), *K. galpinii* (Abyssinia; orange red), *K. leitchlinii* (Abyssinia; red and yellow), *K. pauciflora* (Natal; yellow), *K. uvaria* (South Africa; red and yellow), etc., and many cultivars in variable shades, 'Flaming Torch,' 'Mount Etna,' 'Old Ivory,' 'Orange Glow,' 'Royal Standard,' etc.

KOCHIA   (Summer Cypress)

FAMILY: *Chenopodiaceae*

Named after W. D. J. Koch, a late eighteenth and early nineteenth century German botanist. A small genus from Europe. Only one annual species is cultivated for its compact shape and feathery foliage which changes from bright green to crimson in late summer. A good spot or accent plant.

K. scoparia and variety *trichophylla*.

LACTUCA   is lettuce. There are some flowering species but they are of little value.

LAPEYROUSIA or LAPEIROUSIA

FAMILY: *Iridaceae*

Commemorating J. F. G. de la Peyrouse, a French navigator of the eighteenth century. Small bulbous plants from tropical and South Africa. There is one hardy species which is useful for selected positions on the rock garden.

L. cruenta (syn. *Anomatheca cruenta;* crimson).

LASTHENIA

FAMILY: *Compositae*

The name is supposed to be that of one of Plato's pretty girl pupils. Only one species in cultivation; an annual from California.

L. glabrata (yellow).

LATHYRUS

FAMILY: *Leguminosae*

The ancient Greek name for the pea. Perennial and annual trailing and climbing plants from temperate zones and tropical mountains. The Sweet pea, derived from *L. odoratus,* needs no description, but some of the species deserve to be better known. The garden or culinary pea is *Pisum sativum.*

Perennials: *L. grandiflorus* (Southern Europe; rosy carmine), *L. latifolius* (Everlasting pea; Europe; variable), *L. magellanicus* (Lord Anson's pea; Straits of Magellan; bluish purple), *L. pubescens* (Chile; pale blue), *L. splendens* (California; carmine), etc.

Annuals: *L. azureus* (Southern Europe; blue), *L. odoratus* (Sweet pea; Italy; variable), *L. tingitanus* (Tangier pea; North Africa; purple and red).   *131*

For the many varieties and strains of the modern Sweet pea see seedsmen's lists.

LAVANDULA   (Lavender)

FAMILY: *Labiatae*

From the Latin *lavo*, to wash; since the Greeks and Romans used lavender in their baths. About 20 species of shrubs, sub-shrubs and perennials from the Mediterranean area, the Canary Isles and India. *L. spica* and *L. vera* yield the essential oil and perfume, and are the most commonly grown. Many of the other species are only hardy in the mildest areas.

L. spica (syn. *L. officinalis*; Common lavender; Mediterranean regions; purple) with varieties *alba* (white) and *rosea* (lilac pink), *L. vera* (Dutch lavender; Southern Europe; silvery leaves, pale blue), etc. There are numerous cultivars.

LAVATERA

FAMILY: *Malvaceae*

Commemorating a seventeenth century Swiss naturalist, J. K. Lavater. A genus of some 20 widely distributed annuals, biennials, perennials and shrubs although comparatively few are in cultivation. They are useful for hot, dry positions and the back border.

Perennials: *L. cashmeriana* (India; pale rose), *L. olbia* (Southern Europe; reddish purple).

Biennial: *L. arborea* (Tree mallow; Europe; pale purple) and variety *variegata*.

Annual: *L. trimestris* (Tree mallow; Europe; pink) with variety *alba* (white) and cultivars 'Loveliness,' 'Sunset,' etc.

LAYIA

FAMILY: *Compositae*

Named after G. T. Lay, a nineteenth century naturalist. A small genus of annuals from California. There are several species but only one appears usually to be offered by seedsmen.

L. elegans (Tidy tips; yellow with white tips to the petals).

LEONTOPODIUM

FAMILY: *Compositae*

From the Greek *leon,* a lion, and *pous,* a foot; the shape of the flowers being thought to resemble a lion's paw. A few species of perennials distributed from China to the Andes and to the European Alps. *L. alpinum* is "Edelweiss" of nineteenth century romance, supposed to grow only in the most dangerous and inaccessible places; in fact it will flourish anywhere in dry soil. Reginald Farrer refers to it disparagingly as 'flannel flower.' All have insignificant flowers and whitish bracts.

L. alpinum (Edelweiss; European Alps), *L. japonicum* (China and Japan), etc.

LEPTOSIPHON   See GILIA

LEPTOSYNE

FAMILY: *Compositae*

From the Greek *leptos,* slender; describing the growth of these plants. A small genus of annuals and perennials closely related to Coreopsis; useful for the border and as cut flowers. All come from America and all are yellow.

Perennial: *L. maritima.*

Annuals: *L. calliopsidea, L. douglasii, L. stillmanii.*   133

LEUCOJUM

FAMILY: *Amaryllidaceae*

From the Greek *leukos*, white, and *ion*, a violet; describing the color and scent of the flower. A small genus of bulbous plants from Central Europe and the Mediterranean regions. They are useful for clumps in light shade.

L. aestivum (Summer snowflake; white, tipped green), *L. autumnale* (white flushed pink), *L. vernum* (white).

LEWISIA

FAMILY: *Portulacaceae*

Commemorating Captain Meriwether Lewis, 1774–1809, of the Lewis and Clark expedition. (See also *Clarkia*) Small perennials for the rock garden from Western America.

L. brachycalyx (white or pink) and variety *angustifolia*, *L. howellii* (apricot pink), *L. rediviva* (soft pink), *L. tweedyi* (flesh pink), etc., and many hybrids and named cultivars.

LIATRIS

FAMILY: *Compositae*

Derivation unknown. Handsome tall perennials for the border from North America.

L. elegans (white), *L. pycnostachya* (Kansas gayfeather, Blazing star; purple), *L. spicata* (purple), etc.

LIGULARIA

FAMILY: *Compositae*

From the Latin *ligula*, a strap; describing the long narrow ray florets. About 80 species of perennials, mostly from China and Japan; once included in *Senecio* but now placed in their own genus. They are tall perennials best in slight shade and moist situations.

L. dentata (syn. *L. clivorum;* China; orange yellow) and cultivars 'Desdemona,' 'Gregynog Gold,' etc., *L. japonica* (Japan; yellow), *L. veitchiana* (China; yellow), *L. wilsoniana* (China; golden yellow), etc.

LILIUM

FAMILY: *Liliaceae*

The Persian word *laleh* and the Greek *leiron* both mean "Lily" and are thought to apply to *L. candidum* whose cultivation goes back for more than three thousand years in parts of the Old World; the Madonna lily of Christian symbolism, but which in fact can be seen depicted in ancient Cretan wall paintings dated far earlier. The genus comprises 85 species or more distributed all round the northern temperate hemisphere and of these approximately 49 are Asiatic, 24 North American, 10 European, and 2 Eurasian in origin. These are now arranged in a number of divisions which are too complex to treat here. For most gardeners today the modern hybrids are replacing many of the species and we owe much to the great hybridisers—especially to de Graaf of Oregon—who have transformed the lily from a subject for specialists to a comparatively easy garden plant while still retaining and even enhancing its beauty. Nevertheless the species should not be neglected and the following is a much abbreviated list of those which are by no means difficult, providing some attention is given to their relatively modest needs.

L. amabile (Korea; grenadine red), *L. auratum* (Gold band or Golden rayed lily; Japan; white and yellow), *L. canadense* (Meadow lily; Eastern North America; pale orange to yellow), *L. candidum* (Madonna lily; probably Mediterranean regions; white), *L. hansonii* (Korea; orange), *L. henryi* (China; deep yellow), *L. pardalinum* (Western tiger lily; Western North America; orange), *L. regale* (Regal lily; Western China; white flushed pink to maroon on reverse), *L. speciosum* and varieties (Japan; variable white to suffused pink), *L. superbum* (American turk's cap lily; Eastern North America; deep yellow to orange red). There are many varieties and many more hybrids and cultivars; see specialist publications—of the New York State Agricultural Experiment Station, 'Lilies of the World' by H. B. Drysdale Woodcock and W. T. Stearn, etc.,—and bulbsmen's catalogs.

LIMNANTHES

FAMILY: *Limnanthaceae*

From the Greek *limne,* a marsh, and *anthos,* a flower; referring to the liking these plants have for damp ground. Showy annuals from California, very attractive to bees. Only one species is cultivated.

L. douglasii (Butter and eggs; yellow and white).

LIMONIUM

FAMILY: *Plumbaginaceae*

From the Greek *leimon,* a meadow; because some species are found growing in salt marshes. Annuals, perennials and sub-shrubs from all parts of the world particularly near the coasts, formerly known as *Statice,* and still sometimes so listed. The branching spikes of flowers are often dried and used for winter decoration. Some species are for the greenhouse but these following are hardy in most areas. See also *Armeria.*

Perennials: *L. bellidifolium* (Europe; white or blue), *L. incanum* (Siberia; crimson) and varieties, *L. latifolium* (South Russia; blue), *L. sinense* (China; yellow), *L. vulgare* (syn. *Statice limonium*; Sea lavender; Europe; purple), etc.

Annuals: *L. sinuatum* (strictly a perennial but usually grown as annual; Mediterranean regions; blue and cream) and cultivars 'New Art Shades,' 'Chamois Rose,' 'Pacific Giants,' etc., *L. spicatum* (Western Asia; rose or white), etc.

LINARIA   (Toadflax)

FAMILY: *Scrophulariaceae*

From the Latin *linum,* flax; referring to the flax-like leaves. Mostly annuals and perennials from the Northern Hemisphere. They are cheerful plants for a vacant spot in the garden; *L. alpina* will scramble about gaily on the rock garden.

Perennials, for the border: *L. dalmatica* (Southeast Europe; yellow, purple and blue), *L. macedonica* (Macedonia; yellow), *L. purpurea* (Southern Europe; purple), *L. vulgaris* (Britain; yellow), etc.

Perennial for the rock garden: *L. alpina* (European Alps; purple, blue and yellow).

Annuals for massing: *L. maroccana* (Morocco;   137

violet purple) and cultivars 'Fairy Bouquet,' 'Northern Lights,' etc., *L. reticulata* (Portugal; purple and yellow), etc.

LINNAEA

FAMILY: *Caprifoliaceae*

Commemorating Carolus Linnaeus, the Swedish botanist, since *L. borealis* was one of his favorite plants. A small genus (according to some botanists only one species) which is found as far north as Alaska. The single species in cultivation is an attractive little trailing evergreen for a shady, moist situation in the rock garden.

L. borealis (soft pink) and variety *americana.*

LINUM   (Flax)

FAMILY: *Linaceae*

From the old Greek name *linon,* used by Theophrastus. An important genus which contains not only one of the world's most important economic and historic plants—*L. usitatissimum* which supplies flax fibre for linen and linseed oil, and was probably first cultivated in Mesopotamia four thousand or more years ago—but a number of extremely attractive garden annuals and perennials including some choice plants for the rock garden. It is widely distributed throughout the temperate regions of the world.

Perennials, for the border: *L. flavum* (Golden flax; Northern Europe; yellow), *L. hirsutum* (Europe, Asia Minor; blue and white), *L. narbonense* (Southern Europe; blue), *L. perenne* (Europe; blue), etc.

For the rock garden: *L. alpinum* (European Alps;

blue), *L. capitatum* (Asia Minor; yellow), *L. salsoloides* (Southern Europe; pink), etc.

Annuals: *L. grandiflorum* (North Africa; rose) with variety *coccineum* (rose crimson) and cultivars 'Bright Eyes,' 'Venice Red,' etc., *L. usitatissimum* (Common flax; origin lost; blue), etc.

LITHOSPERMUM

FAMILY: *Boraginaceae*

From the Greek *lithos*, a stone, and *sperma*, a seed; in reference to the stony hard seeds. A genus of about 40 species of biennials, perennials and sub-shrubs; mostly from the Mediterranean regions, but with one from North America. Those in cultivation are trailing evergreens for the sunny rock garden.

L. canescens (Puccoon, Red root; North America; yellow), *L. diffusum* (syn. *L. postratum;* Southern Europe; blue) with cultivars 'Grace Ward,' 'Heavenly Blue,' etc., *L. purpureo-caeruleum* (Britain; blue), etc.

LOBELIA

FAMILY: *Campanulaceae*

Commemorating Matthias de l'Obel, a Flemish botanist and physician to James I of England. A widely distributed genus of some 200 species of annuals, perennials and sub-shrubs. None of those in cultivation is quite hardy except in the mildest areas, but they remain among the most popular of garden plants whether as tall perennials for the border or dwarf annuals for bedding.

Perennials: *L. cardinalis* (Cardinal flower; North 139

America; scarlet) with variety *alba* (white) cultivar 'Queen Victoria,' *L. fulgens,* (Mexico; scarlet), *L. syphilitica* (Eastern United States; purplish blue), *L. tupa* (Chile; scarlet), etc.

Annual and dwarf: *L. erinus* (South Africa; blue) and cultivars 'Blue Gown,' 'Crystal Palace,' 'Prima Donna,' etc. (variable blue to carmine), *L. tenuior* (Western Australia; blue) and cultivars 'Hamburgia,' 'Sapphire,' etc.

LOBULARIA   (Sweet Alyssum or Sweet Alison)

FAMILY: *Cruciferae*

Thought, somewhat obscurely, to be from the Greek *lobulus,* a little lobe. A small genus related to the true *Alyssum.* There are four species, all natives of the Mediterranean regions, but only one usually in cultivation; the small plant much used for edging—usually planted alternately with lobelia—whose supreme dullness is only equalled by its supreme popularity.

L. maritima (syn. *Alyssum maritimum*) with several varieties and variable cultivars, 'Lilac Queen,' 'Rosie O'Day,' 'Violet Queen,' 'Royal Carpet,' etc.

LOPEZIA

FAMILY: *Onagraceae*

Named after Thomas Lopez, a Spanish botanist of the sixteenth century who wrote on the flora of America. A genus of 15 to 20 species of annuals, perennials and shrubs from Central America. Very few are in cultivation and most of these are greenhouse subjects; the one remaining is an annual best treated as half hardy in most areas and of interest mainly for the curious construction of its flowers.

L. coronata (red).

LUNARIA   (Honesty; Moonwort)

FAMILY: *Cruciferae*

From the Latin *luna*, the moon; describing the round and silvery seed cases. Two species from Europe; one of which, the biennial *L. annua*, has managed to escape from gardens since its introduction to Britain 400 years ago and established itself as a wild plant. That is much how it looks in the border, but its sprays of dried seed heads are used for winter decoration.

Perennial: *L. rediviva* (purple).

Biennial: *L. annua* (syn. *L. biennis;* white to pink and purple) with variety *variegata* and cultivar 'Munstead Purple.'

LUPINUS

FAMILY: *Leguminosae*

From the Latin *lupus*, a wolf or destroyer because it was thought that these plants depleted the fertility of the soil by their numbers and strong growth. Over 300 species of annuals, perennials and subshrubs mainly from North America, though there are a few from the Mediterranean area which have been known since Roman times. The Russell hybrids and others, derived mainly from *L. arboreus* and *L. polyphyllus*, are among the showiest of all perennials with a color range embracing almost the entire spectrum, but some of the species are still of interest.

Perennials: *L. nootkatensis* (North-west America; blue, purple and yellow), *L. polyphyllus* (California; blue, white and pink), etc.

Shrubby: *L. arboreus* (California; yellow, white and violet), etc.

Annuals: *L. densiflorus* (California; yellow), *L. hartwegii* (Mexico; blue, white and red), *L. hirsutus* (Mediterranean regions; blue and white), *L. luteus* (Southern Europe; yellow), etc. There are many seedsmen's specialities from these including dwarf strains.

See seedsmen's and nurserymen's lists.

LYCHNIS

FAMILY: *Caryophyllaceae*

From the Greek *lychnos*, a lamp; describing the glowing flowers. A small genus from the north temperate zones of Europe and Asia which contains hardy perennials for the border and rock garden and one good annual. It may be noticed that judging by the number of synonyms here this also is a somewhat restless and uncertain genus, but the following are the names under which these species should now be seen listed.

Perennials for the border: *L. chalcedonica* (Jerusalem Cross; Maltese cross; Russia; scarlet), *L. coronaria* (syn. *Agrostemma coronaria*; Mullein pink, Rose campion; Europe; magenta), *L. flos-jovis* (syn. *Agrostemma flos-jovis*; Flower of Jove; Europe; bright pink), *L. grandiflora* (Japan; salmon), *L. viscaria* (syn. *Viscaria vulgaris*; German catchfly; Europe; purple), etc.

142

For the rock garden: *L. alpina* (syn. *Viscaria alpina,* Europe; pink), *L. fulgens* (Siberia; vermilion), etc.

Annual: *L. coeli-rosa* (syn. *Silene coeli-rosa;* Rose of Heaven; Eastern Mediterranean; variable purple).

LYCORIS

FAMILY: *Amaryllidaceae*

Said to be named after a beautiful actress in Ancient Rome, an earlier mistress of Mark Antony. A small genus of bulbous plants from Japan and China and related to *Amaryllis belladonna.* Most are greenhouse subjects and there appears to be some doubt which are the hardiest, but these following may be tried outdoors, given a selected site in milder winter areas.

L. radiata (syn. *Amaryllis radiata;* red), *L. sanguinea* (red), *L. squamigera* (syn. *Amaryllis hallii;* rosy lilac).

LYSIMACHIA

FAMILY: *Primulaceae*

From the Greek *luo,* to loose, and *mache,* strife, hence the common name of *L. vulgaris.* The Romans are said to have placed these flowers under the yokes of oxen since they were supposed to keep away flies and gnats and thus relieve the animals from irritation. About 120 species from temperate and sub-tropical regions of the world, of which most in cultivation are hardy perennials and generally rather commonplace. For the border and ground cover.

For the border: *L. atro-purpurea* (Greece; purple),

L. *ephemerum* (Europe; white), *L. vulgaris* (Yellow loosestrife; Britain; yellow), etc.

For ground cover, trailing: *L. nummularia* (Creeping Jenny; Britain; yellow).

LYTHRUM

FAMILY: *Lythraceae*

From the Greek *lythron,* blood; an allusion to the color of the flowers or to the reputed styptic qualities of some species. The common name, "Purple loosestrife," may allude to the same qualities as *Lysimachia* or to an apparent resemblance, apart from color, to *Lysimachia ephemerum.* A small genus of perennials; useful for damp places in the garden.

L. alatum (North America; purple), *L. salicaria* (Purple loosestrife; Britain; reddish purple) with varieties, and cultivars 'Brightness,' 'The Beacon,' etc.

MACLEAYA   (Plume Poppy)

FAMILY: *Papaveraceae*

Commemorating Alexander Macleay, a Secretary of the Linnean Society in the mid-nineteenth century. Two perennials from China often listed as *Bocconia.* Tall and impressive plants for the larger garden. The common name is somewhat misleading since their plumes of minute flowers bear no resemblance whatever to a poppy.

M. cordata (syn. *Bocconia cordata;* pinkish buff), *M. microcarpa* (syn. *Bocconia microcarpa;* yellowish), and cultivar 'Kelway's Coral Plume' (Coral pink).

MAIANTHEMUM

FAMILY: *Liliaceae*

From Maia, in Greek mythology the mother of Mercury, to whom the month of May was dedicated—when these plants flower. Three species of hardy perennials; modest but attractive little plants which may be left to colonise shady places. All are white.

M. bifolium (syn. *Smilacina bifolia;* Northern Europe), *M. canadense* (Canada), *M. dilatum* (California to Alaska and Japan).

MAJORANA

is a genus of which most of the garden forms are culinary herbs; sometimes included under *Origanum.*

MALCOMIA

FAMILY: *Cruciferae*

Named after William Malcolm, a seventeenth century nurseryman and botanist. Though there are 35 species in this genus only one is commonly grown; this is probably the easiest of all annuals and ideal for starting off children on their own gardens. The common name appears to be a misnomer since it is in fact a native of Southern Europe.

M. maritima and varieties (Virginian stock; variable white to pink and mauve).

MALOPE

FAMILY: *Malvaceae*

The old Greek name for a kind of mallow; from a word meaning soft or soothing; see below. A   145 small genus of annuals from the Mediterranean region.

M. malacoides (pink and purple), *M. trifida* (purple) with varieties *alba* (white), *grandiflora* (rosy purple) and *rosea* (rose).

MALVA   (Mallow)

FAMILY: *Malvaceae*

From the Greek *malakos*, soft or soothing; perhaps referring to an ointment made at one time from the seeds. Annual, biennial and perennial plants for the border, all from Europe.

Perennials: *M. alcea* variety *fastigiata* (red), *M. moschata* (Musk mallow; rose) with variety *alba* (white).

Annuals or treated as such: *M. alcea* (naturally a perennial; purple), *M. crispa* (white and purple), *M. sylvestris* (naturally a biennial; purple rose), etc.

Both "Mallow" and "Musk mallow" are sometimes applied locally to other genera or their species in the family *Malvaceae*.

MANDRAGORA

FAMILY: *Solanaceae*

From the Greek *mandragorus*, a herb possessing narcotic properties. Of interest chiefly on account of the many legends attaching to one species. Among other magical powers it was thought that the juice of mandrake would induce fecundity in women and also that it was an infallible love potion. Unfortunately however, the mere act of removing it from the ground was fatal so an animal, usually a dog, was used. He was tied to the plant by a length of cord—at midnight—and offered meat just beyond his reach, thus dragging out the root in his attempts to reach the bait. The dog invariably died, but otherwise the mandrake's precious juices could then be extracted safely. There are two species, both from Southern Europe.

M. autumnalis (violet), *M. officinarum* (Devil's apple, Mandrake; greenish yellow).

MARTYNIA

FAMILY: *Martyniaceae*

Commemorating John Martyn, a British botanist of the eighteenth century. A small genus from Central America and the West Indies. Only one annual species appears to be in cultivation; an attractive plant for the sheltered border best treated as half hardy in most areas.

M. louisiana (syn. *M. annua;* purplish pink marked with yellow, violet and white).

MATRICARIA

FAMILY: *Compositae*

From the Latin *matrix,* the womb; the plant once being used as a cure for female disorders. A genus of about 20 species which is not very clearly distinguished from *Chrysanthemum* or *Anthemis,* but the one perennial usually grown is a useful garden plant and good for cutting.

M. maritima (syn. *M. inodora;* White mayweed; Europe).

MATTHIOLA

FAMILY: *Cruciferae*

Named after Pierandrea Mattioli (or Piero Antonio Matthioli) a sixteenth century Italian physician *147* and botanist. About 50 species from the Mediterranean areas and South Africa, of which the few in cultivation are important as being known collectively as "Stocks"; two in particular having given rise to the wide range of popular annual and biennial bedding plants under that name.

Biennials: *M. incana* (Southern Europe; purple) with the 'Brompton,' 'East Lothian,' 'Giant Excelsior' strains, etc. (variable, white to purple and crimson).

Annual: *M. bicornis* (Night-scented stock; Greece; lilac).

Annuals best treated as half hardy: *M. incana* variety *annua* with 'Beauty of Nice,' 'Early Giant Imperial,' 'Ten Week,' 'Trysomic Seven Week,' etc., in named varieties (variable white and yellow to rose and purple). See seedsmen's catalogs.

M. bicornis is sometimes listed as *Hesperis tristis.*

MAZUS

FAMILY: *Scrophulariaceae*

From the Greek *mazos,* a teat, referring to the tubercles at the mouth of the flowers. A small genus related to *Mimulus* and containing a few species of dwarf perennials for the sunny rock garden.

M. pumilio (New Zealand and Australia; purplish blue), *M. reptans* (Himalayas; rosy lavender), etc.

MECONOPSIS

FAMILY: *Papaveraceae*

From the Greek *mekon,* a poppy, and *opsis,* like.

148   Annual, biennial and perennial plants most of

which have the reputation of being difficult but which more than repay the effort to grow them. One of the several mysteries surrounding this beautiful genus is that all the species known to botanists come from the Himalaya-Sikkim-Western China regions or Central Asia with the single exception of *Meconopsis cambrica*, the "Welsh poppy"; and this is an isolated native of the west coast of Britain and parts of Western Europe. *M. cambrica* will grow cheerfully almost anywhere and *M. betonicifolia* is probably the easiest of the Asian species, but any of the following are worth trying and are not impossible in cool moist areas.

Perennial: *M. cambrica* (Welsh poppy; Britain and Europe; yellow) and varieties, *M. grandis* (Sikkim; variable blue to violet), *M. nepaulensis* (syn. *M. wallichii*; Himalayas; variable blue), *M. regia* (Nepal; yellow), etc.

Monocarpic, flowering once only: *M. betonicifolia* (syn. *M. baileyi*; Blue poppy; Himalayas; azure blue), *M. horridula* (Central Asia; variable blue, sometimes to claret), *M. paniculata* (Western China; yellow), *M. superba* (Bhutan, Tibet; pure white), etc.

It should be noted however that like much else about this genus the terms perennial and monocarpic are variable; some which are supposed to be perennial will die after flowering while others usually described as monocarpic will occasionally survive for several seasons; hence, probably, a certain disagreement in nurserymen's lists.   *149*

MEDEOLA

FAMILY: *Liliaceae*

Named after Medea, the witch princess of Colchis. A small genus of only one species related to *Trillium* and coming from North America. Useful for the rock garden. *M. virginica* (yellow).

MELISSA   is a European genus of culinary herbs. It was at one time considered to be a cure for nervous troubles and melancholy; hence its common name, "Balm."

MELITTIS

FAMILY: *Labiatae*

From the Greek *melissa*, a bee, since the flowers are very attractive to bees. A genus of one perennial from Europe, of little interest except perhaps for the wild garden.

M. melissophyllum (Bastard balm; white).

MENTHA

FAMILY: *Labiatae*

Again a genus of culinary herbs named after another unfortunate Greek nymph, Mentha, who got herself turned into a plant. Most of the species can become a nuisance in the garden but there is one minute and attractive carpeting plant for a cool, moist place in the rock garden.

M. requienii (Corsican mint; mauve).

MENTZELIA

FAMILY: *Loasaceae*

Named after Christian Mentzel, a German botanist of the seventeenth century. About 50 species
150   of annual, biennial and perennial plants from

North and South America. Few are in cultivation and most of these are best in the greenhouse, but there is one attractive annual.

M. lindleyi (syn. *Bartonia aurea;* California; bright yellow).

MERTENSIA

FAMILY: *Boraginaceae*

Named after Francis Karl Mertens, Professor of Botany at Bremen in the early nineteenth century. Attractive low growing perennials for the shaded rock garden or light woodland. *M. ciliata* (North America; pink to blue), *M. echioides* (Himalayas; dark blue), *M. maritima* (Oyster plant; Britain, china blue), *M. virginica* (Virginia bluebells in America, Virginian cowslip in England; North America; purplish to blue), etc.

MESEMBRYANTHEMUM is a genus of greenhouse succulents closely related to and sometimes confused with several other genera. Those mainly used as half hardy annuals for summer bedding are now included in *Dorotheanthus* although most nurserymen still list them under the more commonly accepted name.

MICHAUXIA

FAMILY: *Campanulaceae*

Commemorating André Michaux, an eighteenth century French botanist. A small genus of tall perennials, usually treated as biennials. Now somewhat rare; only one species appears sometimes to be obtainable.

M. campanuloides (Asia Minor; waxy white).

MIMULUS

FAMILY: *Scrophulariaceae*

From the Greek *mimo*, ape; because the flowers were thought to look like a monkey's face. Showy annuals and perennials from many parts of the temperate zones, particularly North America, invaluable for bedding, the waterside and the rock garden. *Mimulus moschatus* is the plant which so mysteriously lost its perfume about the end of the nineteenth century.

Perennials: *M. cardinalis* (Cardinal monkey flower; North America; red or red and yellow), *M. cupreus* (North America; coppery red) and cultivars 'Monarch Strain,' 'Red Emperor,' etc., *M. luteus* (Chile; yellow marked red), *M. moschatus* (Monkey musk; North America; yellow), etc.

Perennials for the rock garden: *M. cupreus* cultivar 'Whitecroft Scarlet,' *M. primuloides* (North America; yellow), etc.

Annuals or best treated as annuals: *M. brevipes* (California; yellow), *M. fremontii* (California; crimson), *M. variegatus* (Europe; yellow, variable markings) and cultivars 'Bonfire,' 'Queen's Prize,' etc.

MIRABILIS

FAMILY: *Nyctaginaceae*

From the Latin *mirabilis*, wonderful or admirable. A small genus of annuals and perennials from the warmer regions of America and therefore tender in less favored areas, where all those in cultivation are best treated as half hardy annuals. *M. jalapa* is curious in that it will often bear different colored flowers on the same plant.

M. jalapa (Four o'clock plant; Marvel of Peru; South America; variable yellows to carmine), *M. longiflora* (Mexico; variable), *M. multiflora* (Western America; rosy purple).

MITCHELLA

FAMILY: *Rubiaceae*

Named after an early American botanist, Dr. John Mitchell. Evergreen trailing plants useful for ground cover in shade.

M. repens (Partridge berry; North America; white flowers, scarlet berries) with variety *leucocarpa* (white berries), etc.

MOLUCCELLA

FAMILY: *Labiatae*

The name is taken from the Moluccas islands in the Malay Archipelago from whence one of the species is thought to have come. Moluccellas are useful for floral arrangements by reason of their enlarged, shell-like calyces borne in long spikes. The species usually grown is best treated as a half hardy annual.

M. laevis (Bells of Ireland; Shell flower; Syria and Western Asia; flowers white, calyces pale green).

MONARDA

FAMILY: *Labiatae*

Named after Nicholas Monardes, a sixteenth century Spanish physician and botanist. A small genus of herbs from North America related to *Salvia*. There are annual and perennial species but generally only the perennials are in cultivation.

M. didyma (Bee balm, Oswego tea; red) with variety *alba* (white), *M. fistulosa* (Bergamot; lavender to purple) with variety *violacea* (violet purple), etc., and cultivars 'Cambridge Scarlet,' 'Salmon Queen,' 'Perry's Variety,' 'Prairie Night,' etc.

MONARDELLA

FAMILY: *Labiatae*

The diminutive of *Monarda*, these plants having the general appearance of dwarfs of that genus. Annuals and perennials from California suitable for the sheltered rock garden.

Perennials: *M. macrantha* (orange red to scarlet), *M. villosa* (scarlet).

Annual: *M. candicans* (white).

MONTBRETIA

FAMILY: *Iridaceae*

Named after A. F. Conquebert de Montbrét, a botanist of the French expedition to Egypt in the eighteenth century. The true species are greenhouse plants and now rarely seen and the familiar garden montbretias are hybrids of *Crocosmia*.

MORINA

FAMILY: *Dipsaceae*

Named after the French botanist, Louis Morin, 1636–1715. Uncommon and elegant perennials from Central Asia, but only reliably hardy in mild winter areas.

M. betonicoides (Sikkim; rose), *M. coulteriana* (Himalayas; yellow), *M. longifolia* (Nepal; pink and white), etc.

154

MUSCARI   (Grape hyacinth)

FAMILY: *Liliaceae*

From the Greek *moschos*, musk; describing the scent of some species. Small bulbous plants from the Mediterranean regions. Some botanists have now divided them into three groups, *Moscharia, Leopoldia* and *Botryanthus*, but generally they are all still listed as *Muscari*. There are about 50 species; the following is a selection of those usually offered by bulbsmen.

M. ambrosiacum (cream and lilac), *M. armeniacum* (blue), *M. botryoides* (blue) and variety *album* (white), *M. comosum* (greenish white and purple) variety *monstrosum* (violet blue), *M. neglectum* (dark blue), *M. tubergenianum* (parti colored; light and dark blue), etc.

MYOSOTIDIUM

FAMILY: *Boraginaceae*

The name is taken from *Myosotis* (Forget me not); referring to the flowers which resemble those of that plant. There is only one species, an attractive perennial from the Chatham Islands, for a cool but sheltered position.

M. hortensia (syn. *M. nobile;* blue and white).

MYOSOTIS

FAMILY: *Boraginaceae*

From the Greek *mus*, a mouse, and *otes*, an ear; in reference to the shape of the leaves. About 40 species of annuals, biennials and perennials from the temperate regions, particularly Europe and Australasia. Most of those in cultivation are peren-   155 nials, but their cultivars and seedsmen's strains are usually treated as biennials for summer bedding. A few of the smaller species are good rock garden plants.

For the border or bedding: *M. australis* (New Zealand; yellow, sometimes white), *M. dissitiflora* (Europe; blue), *M. palustris* (Forget me not; Britain; blue), *M. silvatica* (Europe and Northern Asia; blue), etc., and cultivars, together with cultivars of *M. alpestris*, 'Anne Marie Fischer,' 'Blue Bird,' 'Carmine King,' 'Marga Sascher,' 'Rosea,' etc.

For the rock garden: *M. alpestris* (European Alps; azure blue), *M. azorica* (Azores, and not reliably hardy; violet blue), *A. caespitosa* (Europe; sky blue), etc.

MYRRHIS

FAMILY: *Umbelliferae*

From the Greek *myrrha*, fragrant. There is only one species, a fragrant perennial from Europe, once cultivated as a culinary herb but now sometimes grown in the border.

M. odorata (Myrrh, Sweet Cicely; white).

NARCISSUS

FAMILY: *Amaryllidaceae*

From the Greek *narke*, stupor, or *narkeo*, to be stupefied, since some species have narcotic properties. In Greek mythology these plants were consecrated to the Furies, who were said to have used the narcissus to stupefy those whom they wished to destroy. According to legend Narcissus was a beautiful youth who was loved by the nymph, Echo, but repulsed her; to punish him for his cruelty Aphrodite caused him to fall in love with his own image seen in a fountain and at last to die of despair, when he was transformed into a flower growing beside the pool. In more modern terms a genus of 60 or so species from Europe, North Africa and Western Asia, and many thousands of cultivars. Today although 4,000 of them are obsolete there are over 10,000 named varieties in the *Classified List and International Register of Daffodil Names*, published by the Royal Horticultural Society in consultation with the Royal General Bulb Growers' Society of Haarlem, Holland. Many of them are rare but the gardener still has access to some 500 varieties which are commercially cultivated. These have superseded the species from which they are derived, and the very restricted selection given here is of the smaller species suitable for the rock garden and other special situations.

N. bulbocodium (Hoop petticoat daffodil; Southern Europe; yellow), *N. cyclamineus* (Spain; lemon and yellow), *N. minimus* (syn. *N. asturiensis;* Pygmy daffodil; Europe; yellow), *N. rupicola* (Portugal; deep yellow), *N. triandrus* (Angel's tears; Portugal; white), *N. watieri* (Morocco; white), etc.

The common names "Daffodil" and "Lent lily" probably referred originally to *N. pseudo-narcissus.*

For full classification see publications of daffodil societies in the United States, Britain, Holland, Australia, etc., and bulbsmen's catalogs.

NEMESIA

Family: *Scrophulariaceae*

The ancient Greek name used by Dioscorides for a similar plant. There are some 50 species in this genus; annuals, perennials and sub-shrubs mainly from South Africa, but the garden plants which are usually treated as half hardy annuals for brilliant summer bedding are mostly derived from *N. strumosa.* Colors range almost throughout the entire spectrum.

N. strumosa and cultivars 'Aurora,' 'Blue Gem,' 'Fire King,' etc., strains 'Superbissima Grandiflora,' 'Strumosa Suttonii,' 'Dwarf Triumph Strain,' etc.

NEMOPHILA

Family: *Hydrophyllaceae*

From *nemos,* a glade, and *phileo,* to love; referring to the fact that the plant is found naturally growing in glades or groves. A small genus from North America of which two or three are grown in gardens as easy, attractive annuals.

N. maculata (white, blotched blue), *N. menziesii* (syn. *N. insignis;* Baby blue eyes; light blue) with variety *alba* (white).

NEPETA   (Catmint)

Family: *Labiatae*

An early Latin name, probably from Nepi in Italy. About 150 species of perennials, some of which were once used for their medicinal qualities. A few are now grown for their aromatic foliage and for edging. Their origins appear to be uncertain.

N x *faassenii* (syn. *N. mussinii;* lavender blue), *N. macrantha* (syn. *Dracocephalum sibiricum;* blue), *N. nervosa* (light blue), etc., and cultivars 'Six Hills Giant,' 'Souvenir d'Andre Chaudron,' etc.

NERINE

FAMILY: *Amaryllidaceae*

Probably from the Nereids, the beautiful sea nymphs, daughters of Nereus. A genus of about 30 bulbous plants from South Africa. Nearly all of those in cultivation are for the cool greenhouse but one species may be grown outdoors given a sheltered situation in favorable areas.

N. bowdenii (pink) and cultivars 'Blush Beauty,' 'Pink Triumph,' etc.

NICANDRA

FAMILY: *Solanaceae*

Named after a Greek physician and botanist, Nikander of Colophon, who lived in the second century A.D. There is only one species, an annual from Peru, and its chief claim to distinction appears to be that it is said to be a repellant to white fly if grown near other plants which are liable to this pest.

N. physaloides (Apple of Peru, Shoofly plant; blue and white).

NICOTIANA   (Tobacco plant)

FAMILY: *Solanaceae*

In honor of Jean Nicot, a French consul in Portugal, who introduced the tobacco plant into France and Portugal during the seventeenth century.

There are about 66 species of annual and perennial plants in the genus, 45 from the warmer regions of North and South America, 21 from Australia. The most important economically is *N. tabacum* which provides the tobacco of commerce, although *N. rustica* is still used for this purpose and is said to be the first species to provide smoking tobacco in Europe. The ornamental plants for the border and flower arrangement, most of them highly fragrant, have been largely derived from *N. alata,* the hybrid *N x sanderae* and others. They are usually treated as half hardy annuals.

N. alata with variety *grandiflora* (syn. *N. affinis,* strictly a perennial; Brazil; white), and cultivars 'Daylight,' 'Lime green,' 'Sensation,' etc., *N. x sanderae* (pink to carmine) and cultivars 'Crimson King,' 'Knapton Scarlet,' etc., *N. suaveolens* (South America; white), *N. sylvestris* (Argentina; white), etc.

NIEREMBERGIA   (Cup flower)

FAMILY: *Solanaceae*

Named after Juan Nieremberg, a Spanish naturalist of the first half of the seventeenth century. A genus of 35 species from Mexico and subtropical America. Those in cultivation are of a low, spreading habit and mostly for the greenhouse, but the few which can be grown outdoors in milder districts deserve to be much better known. For a moist, shady position on the rock garden.

Perennial: *N. repens* (syn. *N. rivularis;* South America; white tinged with yellow or pink).

Perennial but best treated as annual: *N. caerulea* (syn. *N. hippomanica;* Argentine; lavender blue).

NIGELLA

FAMILY: *Ranunculaceae*

From the Latin *nigellus,* a diminutive of *niger,* black; referring to the black seeds. A genus of about 20 species, natives of the region stretching from Europe to Eastern Asia, of which only two with their varieties and cultivars are in cultivation. They are all decorative annuals for the border.

N. damascena with varieties *alba* and *flore-pleno* (Love in a mist, Devil in a bush; South Europe; white to deep blue) and cultivars 'Miss Jekyll,' 'Oxford Blue,' 'Persian Jewels,' 'Persian Rose,' etc. (variable including pink), *N. hispanica* (Spain; deep blue).

NOLANA

FAMILY: *Nolanaceae*

From the Latin *nola,* a little bell; referring to the shape of the flowers. About 80 species from South America are known; few if any are in cultivation although the seed of a cultivar is often available. An unusual little annual for the front border or rock garden probably best treated as half hardy in most areas.

N. lanceolata (Peru; blue and white) and probable cultivar 'Lavender Gown.'

NOMOCHARIS

FAMILY: *Liliaceae*

From the Greek *charis,* charm, and *nomos,* pasture; describing the flowers and where they are   *161* found. A genus of 16 species of bulbous plants distributed from the Himalayas to China, and interesting as they show most of the botanical features of both *Fritillaria* and *Lilium*. Beautiful but unfortunately somewhat difficult; they appear to prefer a cool, moist summer climate.

N. aperta (Western China; pink, spotted crimson), *N. farreri* (Burma; pink, spotted red), *N. mairei* (China; white), *N. nana* (Northern India to Tibet; purple-rose), *N. pardanthina* (Yunnan; rosy lilac, blotched maroon), etc.

OCIMUM is a genus of which the garden forms are culinary herbs. The best known is *Ocimum basilicum*, "Sweet Basil."

OENOTHERA

FAMILY: *Onagraceae*

Somewhat fancifully from the Greek *oinos*, wine, and *thera*, pursuing or imbibing; the roots of one (or more probably an allied European plant) being thought to induce a thirst for wine. A genus of 80 species of annuals, biennials and perennials most of them originally from North America, but now naturalised in many parts of the world. The common name, "Evening primrose," relates to the flowers of some of them opening in the evening, *O. biennis* in particular. This is little better than a weed and some of the other species are better for the border. There are several showy plants for the larger rock garden.

Perennials for the border: *O. fruticosa* (Sundrops; lemon yellow) with varieties *major* and *youngii*, *O.*

glaber (golden yellow), *O. perennis* (syn. *O. pumila;* yellow), *O. speciosa* (Mexico; white to pink), etc., and cultivars 'Fireworks,' 'Yellow River,' etc.

For the rock garden: *O. acaulis* (Chile; white turning rose), *O. caespitosa* (California; white), *O. missouriensis* (syn. *O. macrocarpa;* Missouri sundrop; yellow).

Biennials: *O. biennis* (Evening primrose) with varieties *grandiflora* and *lamarckiana,* etc.

OMPHALODES

Family: *Boraginaceae*

From the Greek *omphalos,* navel, and *eidos,* like; referring to the shape of the calyx. A genus of 28 annual and perennial species from the Mediterranean regions, Asia and Mexico, although few are in cultivation. These are suitable for the lightly shaded rock garden.

Perennials: *O. cappadocica* (Turkey; dark blue) and cultivar 'Anthea Bloom,' *O. luciliae* (Greece and Asia Minor; rose, turning blue), *O. verna* (Southern Europe; blue) and variety *alba* (white).

Annual: *O. linifolia* (Venus's navel wort; Southern Europe; white).

ONONIS

Family: *Leguminosae*

The ancient Greek name thought to be from *onos,* an ass, and *onemi,* to delight, since some of the 75 species are good fodder plants. Few of them are of much garden value but there are two small shrubs or sub-shrubs which might find a place in the larger rock garden.

O. *fruticosa* (Europe; purple), O. *rotundifolia* (Europe; rose pink).

ONOPORDON or ONOPORDUM

FAMILY: *Compositae*

From the Greek *onos*, an ass, and *perdo*, to eat, since donkeys are supposed to be partial to the thistle-like foliage. A genus of 45 species of annuals, biennials and perennials from Europe, North Africa and West Africa. Where there is space for them the few in cultivation are tall, stately plants for the border.

Perennials: O. *acanthium* (Scots thistle; Europe and Siberia; purple), O. *arabicum* (North Africa; mauve).

Biennial: O. *tauricum* (Southern Europe; purple).

OPHRYS is a genus of British and European terrestrial orchids, many of them now extremely rare, and remarkable in some cases for the resemblance of their flowers to insects. Since propagation appears to be almost impossible they are seldom, if ever, offered and on no account should any of these plants ever be collected from the wild.

ORCHIS

FAMILY: *Orchidaceae*

From the Greek *orchis*, testicle; in reference to the tuberous root of some species. A genus of about 35 species of terrestrial orchids from Africa, Britain, Europe and North America. A number of the species in this genus also are very rare and again no attempt should be made to collect them.

A few are sometimes offered, occasionally seed, and these are best suited to carefully selected positions in the rock garden.

O. latifolia (Europe; pink), *O. maculata* (syn. *Dactylorchis maculata*; Europe; pink), *O. purpurea* (Britain; purple and pink), *O. spectabilis* (North America; pinkish purple), etc.

ORIGANUM

FAMILY: *Labiatae*

From the Greek *oros,* mountain, and *ganos,* beauty; some species being found in the mountainous regions of the Mediterranean. A genus of hardy and shrubby perennials somewhat confused with *Majorana* which includes several forms of the culinary herb marjoram—"Wild marjoram," *Origanum vulgare,* and "Pot marjoram," *Origanum onites* now known as *Majorana onites,* etc.,—and some decorative plants, listed here, for the rock garden.

O. dictamnus (Cretan dittany; Crete; marbled foliage pink flowers), *O. hybridum* (hybrid; pink), *O. vulgare* variety *aureum* (Europe; yellow foliage and flowers), etc.

ORNITHOGALUM

FAMILY: *Liliaceae*

From the Greek *ornis* or *ornithos,* a bird, and *gala,* milk; 'bird's milk' was supposed to be a colloquial expression among the ancient Greeks for something wonderful. About 150 species of bulbous plants from Southern Europe, South Africa and Asia Minor of which there are relatively few in cultivation. Some of these need greenhouse protection and those listed here are the hardier species.

O. nutans (Southern Europe; silvery white), *O. pyrenaicum* (Southern Europe; pale yellow), *O. umbellatum* (Star of Bethlehem; Europe and North Africa; white), etc. Other species are sometimes offered by bulbsmen.

OSTROWSKIA is a genus of one very handsome species, *O. magnifica*, which unfortunately is rarely seen in cultivation.

OURISIA
FAMILY: *Scrophulariaceae*

Commemorating a one time Governor Ouris, of the Falkland Islands. A small genus, mostly of hardy and attractive little perennials for the rock garden which deserve to be better known. They are natives of the Andes, antarctic parts of South America and New Zealand.

O. alpina (the Andes; pink to red), *O. coccinea* (the Andes; scarlet), *O. elegans* (Chile; red), *O. macrophylla* (New Zealand; white), etc.

OXALIS
FAMILY: *Oxalidaceae*

From the Greek *oxis*, acid; in reference to the sap, possibly of *O. acetosella*, "Wood sorrel." A large and widely distributed genus of annuals and perennials, of which a few perennials are attractive rock garden or front border plants while others can become persistent weeds. Those listed here are hardy and relatively harmless.

O. enneaphylla (Falkland Islands; white flushed pink) with variety *rosea* (pink), *O. magellanica* (Bolivia and Southern Australia; white), *O. oregana* (North America; red), *O. rosea* (syn. *O. floribunda;* Chile; rose and white), etc. Other species are offered by nurserymen from time to time but should be planted with caution.

PAEONIA   (Paeony)

FAMILY: *Paeoniaceae*

Commemorating Paeon, an ancient Greek physician, who is said to have first used *P. officinalis* medicinally. Until recently the genus was considered to be a member of the family *Ranunculaceae* but botanists now place it in a family of its own. A genus of 33 species of herbaceous and shrubby perennials and a few shrubs, some of which have been cultivated in China and Japan for more than a thousand years. The main division of the genus lies between the shrubby and herbaceous species, although botanically the distinctions are much more detailed and complex than that; but in general the tree paeonies are derived from *P. suffruticosa* while most modern hybrids trace back in one way or another to *P. lactiflora* (syn. *P. albiflora*) from Mongolia and China, and the old fashioned garden forms "Double White," "Double rose" and "Double crimson" are the *albo-plena, rosea-plena* and *rubra-plena* varieties of *P. officinalis* from Southern Europe. The following is a much shortened list of the best of the other herbaceous species.

P. anomala (Russia and Central Asia; crimson),

P. broteri (Spain; purplish red), *P. emodi* (Himalayas; white touched with yellow), *P. mlokosewitschii* (Caucasus; yellow), *P. veitchii* (China; purplish red), *P. wittmanniana* (Caucasus; primrose yellow), etc.

PANCRATIUM   is a small genus of bulbous plants allied to *Hymenocallis* and generally only reliable in the greenhouse.

PANSY   See Viola tricolor

PAPAVER

FAMILY: *Papaveraceae*

An ancient Latin plant name of doubtful origin; and a widespread genus of about 100 species of annuals and perennials. Those in cultivation are all good garden plants, and one for the rock garden will naturalise itself freely without becoming a nuisance.

Perennials: *P. atlanticum* (Morocco; orange), *P. nudicaule* (Iceland poppy; sub-arctic regions; variable, white to yellow and orange) and many strains and cultivars, 'Champagne Bubbles,' 'Kelmscott Strain,' 'Red Cardinal,' etc., *P. orientale* (Oriental poppy; Asia Minor; orange scarlet) and cultivars in shades of white to salmon pink, etc.

Perennial for the rock garden: *P. alpinum* (Europe; variable white to orange and salmon).

Annuals: *P. commutatum* (syn. *P. umbrosum*; Caucasus; crimson and black), *P. glaucum* (Tulip poppy; Syria; scarlet), *P. rhoeas* (Corn poppy, Shirley poppy; Europe; scarlet) and strains 'Begonia Flowered,' 'Shirley Reselected,' etc., *P. somniferum*

168

(Opium poppy; Europe and Asia; variable white to purple) and cultivars and strains 'Daneborg,' 'Carnation Flowered,' etc.

PARADISEA   (St. Bruno's lily)

FAMILY: *Liliaceae*

Commemorating a Count Giovanni Paradisi, 1760–1826, of Modena, Italy. Two species of rhizomatous perennials of which only one is in cultivation, an attractive and reliable garden plant best grown in clumps for effect. It is related to *Anthericum* and may sometimes be listed as *Anthericum liliastrum.*

P. liliastrum (Southern European Alps; white).

PARNASSIA

FAMILY: *Saxifragaceae* or *Parnassiaceae*

From Mount Parnassus, the sacred mountain associated with the worship of Apollo and the Muses, from where some of these plants are supposed to have come. In fact the genus, of about 50 species of perennials, is widely distributed. Small plants best suited for the margins of ponds or streams.

P. alpina (Europe; white), *P. californica* (California; white), *P. caroliniana* (North America; white), *P. palustris* (Grass of Parnassus; Northern Hemisphere; white).

PAROCHETUS

FAMILY: *Leguminosae*

From the Greek *para,* near, and *achetos,* a brook; in reference to the plant's liking for moist situations. One species only from the mountains of tropical Africa and Asia. A pretty little perennial carpeting plant for the sheltered rock garden, which has the useful habit of flowering in late summer to autumn and sometimes even on into the winter in mild seasons. It is not reliably hardy in colder areas.

P. communis (blue flushed with rose).

PARONYCHIA   (Nailwort)

FAMILY: *Caryophyllaceae*

From the Greek *paronuchia*, a whitlow, for which the plant was reputed to be a cure, hence the common name "Nailwort." A genus of 50 species, mainly from the Mediterranean area but some from West Africa and America, of which only two are usually seen in cultivation. There are mat forming perennials useful for hot, dry positions on the rock garden.

P. argentea (Southern Europe; silvery bracts), *P. capitata* (syn. *P. nivea;* Mediterranean area; similar).

PARRYA

FAMILY: *Cruciferae*

Commemorating Sir William Edward Parry, an eighteenth century arctic navigator. About 25 species of small hardy perennials from arctic regions or the high mountains of Asia and North America. Little known but attractive plants for a sunny position in the rock garden.

P. arctica (Arctic America; pale purple), *P. integerrima* (Siberia; purple), *P. nudicaulis* (syn. *P. macrocarpa;* Arctic regions; lilac).

PELARGONIUM is strictly a genus of herbaceous and shrubby greenhouse perennials and consists of over 300 known species without the sub species and varieties not yet recorded. Those used for summer bedding known as florist's geraniums, or sometimes described as *Pelargonium hortorum*, are listed under strain or cultivar names. There is a large number of these but probably the most interesting development of recent years is the appearance of the 'Pan American F_1 Hybrid Carefree' strain. These may be quite successfully treated as half hardy annuals. See seedsmen's and nurserymen's catalogs.

PELTIPHYLLUM

FAMILY: *Saxifragaceae*

From the Greek *pelta*, a shield, and *phyllon*, a leaf; referring to the shape of the leaves. There is only one species, from North America; a handsome perennial for planting on the banks of streams and ponds in larger gardens. It was at one time called *Saxifraga peltata*.

P. peltatum (white to pale pink).

PENSTEMON

FAMILY: *Scrophulariaceae*

From the Greek *pente*, five, and *stemon*, stamen, in reference to the five stamens of the flower, sometimes wrongly spelt Pentstemon. A genus of more than 250 species of annuals, perennials and sub-shrubs almost exclusively from North America. The popular bedding penstemons, derived originally from a crossing between *P. cobaea* 171 and *P. hartwegii,* are the best known but several of the species are good border plants and some are excellent for the rock garden.

Perennials: *P. angustifolius* (Western United States; blue), *P. antirrhinoides* (California; lemon yellow), *P. bridgesii* (North America; scarlet), *P. heterophyllus* (California; sky blue), *P. spectabilis* (Mexico and Southern California; rosy purple), etc., and many cultivars 'Cherry Glow,' 'Garnet,' 'Six Hills Hybrid,' etc., *P. barbatus* is now more correctly *Chelone barbata.*

For the rock garden: *P. davidsonii* (California; dark red), *P. menziesii* (North-western America; purple), *P. rupicola* (North-western America; ruby red), etc.

Annuals are sometimes offered by seedsmen as mixed varieties.

PERILLA

FAMILY: *Labiatae*

Probably the native Indian name. A small genus of 4 to 6 species of which only one is sometimes seen in cultivation; a plant much in favour during the Victorian vogue for carpet bedding on account of its richly colored foliage. Best treated as a half hardy annual in most areas.

P. frutescens variety *nankinensis* (China and Northern India; bronze purple leaves).

PETUNIA

FAMILY: *Solanaceae*

From the Brazilian name for tobacco; *petun; petunia* is closely related to *Nicotiana.* A genus of 40

172

species of annuals and perennials from South America which are rarely if ever seen in gardens. The popular bedding plants, in upwards of 400 varieties tested by Pennsylvania State College some years ago, have been derived from only two or three of them; *P. integrifolia* (purple), *P. nyctaginiflora* (white) and *P. violacea* (violet), all from Argentina. Modern cultivars now include many F_1 hybrid strains and named varieties; 'Theodosia,' 'Romany Lass,' 'White Beauty,' 'Giants of California,' 'All Double Multiflora,' 'Cascade Mixed,' etc. See seedsmen's catalogs.

PHACELIA

Family: *Hydrophyllaceae*

From the Greek *phakelos,* a bundle; relating to the clustered arrangement of the flowers. 200 species of annual and perennial plants from the Andes and North America; of which a number of the annuals, listed here, are showy garden plants.

P. campanularia (California; intense blue), *P. congesta* (Texas; lavender), *P. grandiflora* (California; lavender veined violet), *P. minor* (syn. *Whitlavia minor;* Californian bluebell; California; deep violet), *P. viscida* (syn. *Eutoca viscida;* California; blue), and 'Musgrave' strain, etc.

PHLOX

Family: *Polemoniaceae*

From the Greek *phlox,* a flame; referring to the bright colors of the flowers. A genus of nearly 70 species of annuals and perennials, mostly natives of North America and Mexico. The showy and multicolored modern border plants are nearly all derived from *P. paniculata*—which may sometimes be listed as *P* x *decussata*—and the many forms of equally brilliant annuals have been developed from *P. drummondii*. There are also some valuable species for the rock garden.

Perennials for the border: *P. carolina* (Eastern United States; purple to pink and white), *P. glaberrima* (Eastern North America; red), *P. maculata* (Eastern North America; violet and purple) and numerous cultivars from these in the early flowering classes; *P. paniculata* (syn. *P* x *decussata*; North America; violet purple), and cultivars 'Border Gem,' 'Brigadier,' 'Frau Antonin Buchner,' etc.

For the rock garden: *P. amoena* (Southeastern United States; rose), *P. bifida* (Eastern North America; pale violet to white), *P. divaricata* (syn. *P. canadensis*; North America; blue lavender), *P. douglasii* (Western North America; lilac), *P. subulata* (Eastern United States; purple or white).

Annual: *P. drummondii* (Texas and New Mexico; wide color range) and many cultivars and strains best treated as half hardy in most areas, 'Brilliant,' 'Isabellina,' 'Purple King,' 'Gigantea Art Shades,' etc. See nurserymen's and seedsmen's lists.

PHYLLODOCE

FAMILY: *Ericaceae*

The name of a sea nymph in Greek classical literature. A small genus of very dwarf evergreen shrubs from arctic and alpine regions. Some of the species are attractive plants for the cool, shaded rock garden.

P. breweri (California; purplish rose), *P. caerulea* (Northern Europe; purplish blue), *P. nipponica* (Japan; white tinted pink).

PHYSALIS

FAMILY: *Solanaceae*

From the Greek *physa*, a bladder, in reference to the decorative inflated calyx. About 100 species of annual and perennial plants mostly from Mexico and North America, although the best known extends to Southeastern Europe and Japan. The other species are greenhouse subjects, but this is a useful perennial for the border and its sprays of dried calyces are often used for winter decoration. It can become invasive.

P. alkekengi (syns. *P. bunyardii* and *P. franchettii;* Chinese Lantern plant; North America to Japan; orange scarlet calyces) with varieties *gigantea* and *pygmaea.*

PHYSOSTEGIA

FAMILY: *Labiateae*

From the Greek *physa*, a bladder, and *stege*, a covering; a reference to the formation of the calyx. A small genus of interesting hardy perennials from North America; of which only one, with its varieties, is usually found in cultivation. This is known as "Obedience" or the obedient plant because its separate blooms may be moved about the flower spike and will stay where placed for some time.

P. virginiana (syn. *Dracocephalum virginianum;* False dragonhead, Obedience; rosy pink) with varieties *alba* and *grandiflora* and cultivars 'Summer Glow,' 'Rosy Spire,' 'Vivid,' etc.

PHYTEUMA

FAMILY: *Campanulaceae*

The derivation is obscure but possibly from the Greek *phyteuo*, to plant, or *phyton*, a plant; the name was adapted by Linnaeus from Dioscorides. A genus of 40 or so species of perennials from Europe and Asia. The few in cultivation are for the border and rock garden.

For the border: *P. spicatum* (Europe; cream), etc.

For the rock garden: *P. comosum* (Yugoslavia; blue), *P. halleri* (Swiss Alps; violet), *P. nigrum* (Central Europe; blue or dark violet), etc.

PLATYCODON (Balloon flower)

FAMILY: *Campanulaceae*

From the Greek *platys*, broad, and *kodon*, a bell; describing the shape of the flower. A genus of only one species, a useful and beautiful hardy perennial from China and Japan, where this plant has been cultivated for centuries on account of its reputed medicinal properties, while its elegant flower may often be seen used in Japanese fabric designs and silk pictures.

P. grandiflorum (syn. *Campanula grandiflora*; blue) with varieties *album* (white), *praecox* (violet blue), *roseum* (variable pink). *Mariesii* is a dwarf form which is excellent for a selected position in the rock garden.

PLATYSTEMON (Cream cups)

FAMILY: *Papaveraceae*

From the Greek *platys*, broad, and *stemon*, stamen; referring to the form of the flowers. A genus of 60 species (or according to some botanists only one, but this highly variable) from North Western America. An annual which is said to cover acres of the Californian countryside in spring.

P. californicus (cream) with variety *crinitus* (deeper yellow).

PODOPHYLLUM

FAMILY: *Podophyllaceae*

From the Greek *podos,* a foot, and *phyllon,* a leaf; a contraction of *anapodophyllum,* duck's-foot leaved. A genus of 10 species, mostly from Asia, of which two small perennials are sometimes grown, one of them from North America. They are best suited for a moist and shady part of the wild garden. *P. peltatum* is said to be poisonous.

P. emodi (Himalayas; white flowers, coral fruit), *P. peltatum* (May apple; Eastern North America; white flowers, yellow fruits).

POLEMONIUM

FAMILY: *Polemoniaceae*

Somewhat fancifully from the Greek *polemus,* war, referring to the lance shaped leaflets, or possibly because the plant was named by Dioscorides after King Polemon of Pontus. About 50 species of perennials distributed between Eurasia, North America, Mexico and Chile. Some of them are good border plants, including an interesting yellow form, and one or two are easy subjects for the rock garden.

P. caeruleum (Jacob's ladder; Britain; blue), *P. carneum* (Europe and North West America; cream   177 to pink or blue), *P. cashmirianum* (Kashmir; blue), *P. flavum* (New Mexico; yellow), etc.

For the rock garden: *P. confertum* (North West America; blue), *P. reptans* (North America; blue or white), etc.

POLIANTHES

FAMILY: *Amaryllidaceae*

Probably from the Greek *polios*, white or shining, and *anthos*, flower. There are fifteen species in the genus—or according to some botanists merely one with fourteen varieties—mostly from Mexico or Trinidad, but the real origin of the "Tuberose," which has been cultivated for the sake of its powerful scent since remote times, now seems to be unknown. It is strictly a greenhouse plant, but is often used for summer flowering outdoors in suitably mild areas.

P. tuberosa (Tuberose white) and variety *flore-pleno* with cultivars 'The Pearl,' 'Mexican Everblooming,' etc.

POLYANTHUS

FAMILY: *Primulaceae*

From the Greek *polys*, many, and *anthos*, flower; of the clustered flower heads. Botanically polyanthus should be classed with *Primula* of which it is a highly developed garden form with a long history. As far back as 1640 John Parkinson, the famous apothecary, gardener and author of his time, mentioned a red flowered primrose—"John Tradescant's Turkie-purple primrose"—which had recently been introduced from the Caucasus. It would seem that it might have been this "primrose" which, when crossed with the native yellow oxlip, *Primula elatior*, perhaps a hybrid itself (*P. veris* x *P. vulgaris*), was to produce the red polyanthus or "big oxlip" first described by John Rea in his *Flora, Ceres and Pomona* published in 1665. Later developments of the polyanthus, particularly in its gold laced forms, became a florists' subject for the next two hundred years, but it was not until this century that the modern strains of vigorous plants with large flowers in a wide range of pure, clear colors started to appear. These came from Suttons of Reading, Frank Reinelt of California from about 1930 and later, Mrs. Florence Bellis of Oregon from 1935, Blackmore and Langdon of Bath and many others, including the New Zealand plant breeders.

Strains: Barnhaven 'Marine Blues,' 'Grand Canyon,' 'Desert Sunsets,' etc. (U.S.A.), Blackmore and Langdon's 'Blue' and 'Pink' (Britain), Dean's 'Otago Superba' (New Zealand), Harrison's 'Tango Supreme' (New Zealand), Reinelt's 'Pacific Giants' (U.S.A.), Sutton's 'Triumph' (Britain), and many others. See seedsmen's lists.

POLYGALA

FAMILY: *Polygalaceae*

From the Greek *polys,* many or much, and *gala,* milk; since it was thought that the presence of some of the species in pasture increased the yield of milk. A widespread genus of 500 or 600 species of annuals, perennials and shrubs, most of no importance but including some subjects for the greenhouse and several small shrubs, listed here, for the rock garden.

P. calcarea (Europe; purple, rose or blue) and cultivar 'Bulley' (deep blue), *P. paucifolia* (North America; evergreen, carmine), etc.

POLYGONATUM

FAMILY: *Liliaceae*

From the Greek *polys,* may, and *gonu,* a small joint; referring to the jointed rhizomes. A genus of 50 species from the north temperate zones. The plant itself is considerably less interesting than Gerard's remarks about it in the *Herball* (1597). "The root of Solomon's seale stamped (bruised) while it is fresh and greene, and applied, taketh away in one night, or two at the most, any bruise, black or blew spots gotten by falls or women's wilfulness, in stumbling upon their hasty husband's fists, or such like."

P. biflorum (North America; green and white), *P. commutatum* (United States; white), *P. multiflorum* (David's harp; Europe and Asia; white), *P. odoratum* (syn. *P. officinale;* Solomon's seal; Europe; white), etc.

POLYGONUM

FAMILY: *Polygonaceae*

From the Greek *polys,* many, and *gonu,* a small joint; alluding to the many joints in the stems. Another cosmopolitan genus of some 300 species and again with few of any interest. Some of the larger shrubs, particularly the climbing *P. bald-*

schuanicum, can develop into rampant monsters but a few small species are suitable for the rock garden.

P. affine (Himalayas; rose pink), with cultivar 'Darjeeling,' *P. tenuicaule* (Japan; white), *P. vaccini-ifolium* (Himalayas; rose), etc.

PORTULACA   (Purslane)

FAMILY: *Portulacaceae*

Possibly from the Latin *porto,* to carry, and *lac,* milk; of the milky sap. About 200 species of annuals and perennials, widespread in tropical and sub-tropical regions. The most generally grown is a creeping and trailing annual useful for its masses of showy blooms in hot, sunny positions; it is usually treated as half hardy. *P. oleracea* is a herb which is sometimes used for flavoring salads.

P. grandiflora (Brazil; white and yellow to orange and red) and nurserymen's strains, 'Jewel,' 'Improved Double Mixed,' etc., *P. oleracea* (Southern Europe; yellow).

POTENTILLA   (Cinquefoil)

FAMILY: *Rosaceae*

From the Latin *potens,* powerful; some species having useful medicinal properties. A large genus of 500 species, mostly herbaceous perennials with a few shrubs or sub-shrubs and mainly from the northern temperate and arctic regions. With the exception of the larger shrubs nearly all of those in cultivation are for the rock garden or front border. The following is a selection of the best.

For the front border: *P. atro-sanguinea* (Himalayas;   *181* crimson), *P. grandiflora* (European Alps; yellow), *P. nepalensis* (Himalayas; magenta rose), *P. recta* (Europe to Siberia; yellow), etc., with hybrids and cultivars.

For the rock garden: *P. alba* (Europe; white), *P. aurea* (European Alps; yellow), *P. fragiformis* (North Western America; yellow), *P. nepalensis* var *willmottiae* (carmine), *P. nitida* (Europe; rose) with varieties *alba* (white) and *grandiflora* (pink), etc.

PRIMROSE   See Primula

PRIMULA

FAMILY: *Primulaceae*

From the Latin *primus,* first; referring to the early flowering of some of the species, as the primrose. A complex genus of over 500 species widely distributed throughout the northern hemisphere; it has now been divided by botanists into 30 or more divisions—Auricula, Candelabra, Capitatae, Farinosae, etc.,—and some of these subdivided again. It embraces a wide variety from the humble primrose (*P. vulgaris*) and cowslip (*P. veris*) to the stately candelabra types, the showy species for the greenhouse with their hybrids and cultivars, and the rare and difficult treasures which tantalise the specialist. The following selection is of necessity much abbreviated and it is divided simply into taller and dwarf types more suitable for the rock garden; but all of these are reliable garden species provided they are given the one thing which most primulas demand—a cool, moist soil with plenty of leaf mould or peat. The division to which each belongs is bracketed immediately after its name.

Taller species: *P. beesiana* (Candelabra; Yunnan; rosy purple), *P. bulleyana* (Candelabra; Yunnan; orange), *P. denticulata* (Denticulata; Drumstick primula; Himalayas; variable white to lavender and rose), *P. florindae* (Sikkimensis; Tibet; sulphur yellow), *P. japonica* (Candelabra; Japan; purplish red), *P. pulverulenta* (Candelabra; China; wine red), etc., and varieties and cultivars.

For the rock garden: *P. allionii* (Auricula; Maritime Alps; rose to deep red), *P. auricula* (Auricula; European Alps; yellow), *P. capitata* (Capitatae; Tibet; violet), *P. clarkei* (Farinosae; Kashmir; rose pink), *P. frondosa* (Farinosae; Balkans; rosy lilac), *P. juliae* (Vernales; Caucasus; lilac purple), *P. rosea* (Farinosae; Himalayas; brilliant carmine), *P. vulgaris* variety *rubra* (Vernales; Europe; pink), etc., and varieties and cultivars.

Primrose, *P. vulgaris* (syn. *P. acaulis*) is now offered in as many and as brilliant shades as Polyanthus; *Primula juliae* is the parent of the 'Wanda' strain (variable wine red to blue and purple).

See specialist and nurserymen's lists.

PTEROCEPHALUS

FAMILY: *Dipsaceae*

From the Greek *pteron*, a wing, and *cephala*, head; since the flower head is large for the size of the plant and appears to be covered with feathers. A widely distributed genus of 25 species only one of which is found in cultivation; an easily grown and attractive subject for the rock garden. This was a particular favorite of Miss Gertrude Jekyll, the well known garden designer and writer.

P. parnassi (Greece; purplish pink).

PULMONARIA

FAMILY: *Boraginaceae*

From the Latin *pulmo*, lung; possibly because an infusion from one species was considered to be a cure for lung diseases. A genus of 10 species of hardy perennials, all natives of Europe. In some the flowers change from red to blue as they age— hence the old English country name, "Soldiers and sailors"—and in others the leaves are spotted with paler green or white, so called "Spotted dog."

P. angustifolia (Soldiers and sailors; pink changing to blue) with varieties *alba* (white) and 'Mawson's Variety,' *P. officinalis* (Lungwort, Jerusalem cowslip, Spotted dog; pink then violet), *P. rubra* (red), *P. saccharata* (pink), etc.

PULSATILLA

FAMILY: *Ranunculaceae*

The name first given to this plant by Pierandrea Mattioli, a sixteenth century Italian botanist; it is said to mean shaking in the wind. 30 species from the temperate regions of Europe and Asia, closely related to *Anemone*, in which genus it was once included. The cultivated forms are among the most beautiful of plants for chosen points in the rock garden.

P. alpina (European Alps; blue buds opening to white) and variety *sulphurea* (pale yellow), *P. halleri* (European Alps; deep violet), *P. slavica* (Central

Europe; purple), *P. vernalis* (high alpine meadows; purple buds opening to white), *P. vulgaris* (Pasque flower; Europe; purple) with varieties *alba* (white), 'Budapest' (powder blue), and cultivars which sometimes include red flowered seedlings.

PUSCHKINIA   (Lebanon Squill;)
                 (Striped Squill)

FAMILY: *Liliaceae*

Commemorating Count Apollos Mussin-Puschkin, an eighteenth century Russian chemist and plant collector. Two species of small bulbous plants from Western Asia, allied to *Scilla* and *Chionodoxa*, of which only one is likely to be found in cultivation. Useful for the front border or rock garden.

P. scilloides, often seen listed as *P. libanotica* (pale blue with a darker stripe) and variety *alba* (white).

PYRETHRUM

FAMILY: *Compositae*

From the Greek *pyr*, fire; either in reference to the color of the flowers or to the fact that an infusion from one species was once used as a febrifuge. The popular perennial garden plants are descended from another which although long known as "pyrethrum" belongs in fact to *Chrysanthemum*.

P. roseum (*Chrysanthemum coccineum*; Painted daisy; Caucasus and Persia; variable white to red) and many cultivars including 'Avalanche,' 'Bressingham Red,' 'Kelway's Glorious,' 'Pink Bouquet,' etc. See nurserymen's lists.

QUAMOCLIT

FAMILY: *Convolvulaceae*

The derivation is uncertain, but "Quamoclit" is possibly a native Indian name. A genus of twelve species of annual and perennial climbers from America which are included by some botanists in *Ipomoea*. Three species and a hybrid are generally cultivated, and except in the most favorable climates they are best treated as half hardy annuals for a sheltered situation.

Q. coccinea (Tropical America; scarlet and yellow), *Q. lobata* (Mexico; crimson fading to white), *Q. pennata* (Cypress vine; Tropical America; scarlet), *Q* x *sloteri* (Cardinal climber; crimson).

RAMONDA

FAMILY: *Gesneriaceae*

Named after L. F. Ramond, a French botanist and traveller in the late eighteenth century. A few species of European alpine plants, related to *Haberlea* and *Jankaea*. All are attractive subjects for the lightly shaded rock garden.

R. myconii (syn. *R. pyrenaica*; Pyrenees; lavender) with varieties *alba* (white) and *rosea* (pink), *R. nathaliae* (Bulgaria and Serbia; lavender) with variety *alba* (white), *R. serbica* (Balkans; lilac).

RANUNCULUS

FAMILY: *Ranunculaceae*

From the Latin *rana*, a frog; since many species flourish in marshy places where frogs also are found. A genus of 400 species widely distributed from the high Alps to watery meadows, containing a number of fine garden plants and some of the most invasive weeds. Those in cultivation are perennials, and they include a few beautiful subjects for the rock garden with some good border plants. The following are among the best.

For the border: *R. aconitifolius* (Fair maids of Kent; Europe; white) with variety *flore-pleno*, *R. asiaticus* (the florist's "Turban" or "Persian" ranunculus; Asia; variable white and yellow to crimson) and many strains 'Florentine,' 'Grandiflora,' 'Peony flowered,' etc., see bulbsmen's catalogs, *R. lingua* (Europe; yellow; for a damp situation), *R. lyallii* (New Zealand; white), etc.

For the rock garden: *R. alpestris* (European Alps; white), *R. bullatus* (Mediterranean regions; yellow to orange), *R. ficaria* variety *aurantiacus* (Celandine; Europe; coppery yellow) and cultivar 'Primrose,' *R. glacialis* (European mountains and arctic; white tinged rose), etc.

The common Buttercup is *Ranunculus acris.*

RAOULIA

FAMILY: *Compositae*

Commemorating Edouard Raoul, a French naval surgeon of the first half of the nineteenth century who collected and wrote about plants from New Zealand. A genus of 25 species which includes a few very small, close growing carpeting plants for the rock garden. None of the following reaches more than a half inch in height.

R. australis (silvery foliage), *R. glabra* (emerald), *R. tenuicaulis* (grey green; yellow flowers), etc.

REINECKIA

FAMILY: *Liliaceae*

Named after J. Reinecke, an eighteenth century German gardener and specialist in tropical plants. A genus of perennials containing only one species; a native of China and Japan and hardy in situations not exposed to extreme winter cold. It is attractive as edging or for the rock garden.

R. carnea (pink; mottled yellow throat) and variety *variegata* (cream variegated leaves).

RESEDA

FAMILY: *Resedaceae*

From *resedo*, to heal, the name given by Pliny to one species which was said to have the property of healing external bruises. A genus of some 60 species of annual and biennial plants from the Mediterranean regions, of which the few in cultivation are grown mainly for their fragrance.

Biennial: *R. alba* (Southern Europe; white).

Annual: *R. odorata* (Mignonette; North Africa, Egypt; yellowish white) and cultivars 'Goliath,' 'Machet,' 'Golden Queen,' 'Red Monarch,' etc.

RHAZIA

FAMILY: *Apocynaceae*

Commemorating an Arabian physician and author of medical works in the tenth century, Bekr-el-Rasi. Two species of perennials from the Levant and Asia Minor. Only one is likely to be seen in cultivation and is an unspectacular but attractive border plant.

R. orientalis (lavender blue).

RHEUM   (Rhubarb)

FAMILY: *Polygonaceae*

From the Greek *rha* or *rheon*, rhubarb. A genus of about 50 species from temperate and sub-tropical Asia, which includes culinary rhubarb and a few large decorative perennials for the wild garden where space is no consideration.

R. alexandrae (China and Tibet; yellow bracts), *R. nobile* (Himalayas; pale yellow bracts), *R. officinale* (Tibet; pinkish white or green), *R. palmatum* (China; cream to deep red), etc.

RHODODENDRON   is a genus of generally large shrubs.

RHODOHYPOXIS

FAMILY: *Hypoxidaceae*

From the Greek *rhodon*, rose, and *Hypoxis*, referring to the close affinity of these plants to that genus. Two species of tuberous rooted perennials from the mountains of South Africa, of which only one is in general cultivation. A tiny but charming plant for the sheltered rock garden.

R. baurei (rose) with variety *platypetala* (white) and cultivar 'Apple Blossom.'

RODGERSIA

FAMILY: *Saxifragaceae*

Named after Admiral John Rodgers of the United States Navy in the mid-nineteenth century and commander of an expedition during which *Rodgersia podophylla* was discovered. A few species of handsome hardy perennials from China and

Japan which are useful for waterside or wild gardens.

R. aesculifolia (pinkish white), *R. pinnata* (pink) with varieties *alba* (white) and *superba* (deep pink), *R. podophylla* (buff-yellow), *R. sambucifolia* (white), etc.

ROMANZOFFIA

FAMILY: *Hydrophyllaceae*

Named after Count Nicholai von Romanzoff, a patron of science who organised a voyage round the world in the early nineteenth century. A small genus of dwarf perennials resembling saxifrage found from Alaska to California and possibly in Siberia. Suitable for the rock garden.

R. californica (Western North America; white), *R. unalaschkensis* (Alaska and Canada; pale mauve).

ROMNEYA

FAMILY: *Papaveraceae*

Named after the Rev. T. Romney Robinson, a nineteenth century Irish astronomer, who discovered *Romneya coulteri*. Two species of large semi-shrubby perennials from California, good in milder winter areas and where there is ample space for them to spread.

R. coulteri (Californian tree poppy; white), *R. trichocalyx* (white).

ROSA (Rose)

FAMILY: *Rosaceae*

A very old name of uncertain derivation probably alluding to the typical flower color, perhaps from the Celtic *rhod*, red. Few plants have more history and legends attached to them but probably the most charming is the account of Sir John Mandeville—a great teller of tall stories in the fourteenth century—who relates that a Jewish maid of Bethlehem was loved by one Hamuel, a brutish sot; but she rejected his advances and in revenge he denounced her as a witch and she was condemned to be burned at the stake. God however averted the flames, the stake itself budded and the maiden was seen to stand unharmed under a rose tree full of red and white blossoms; "The first on earth since Paradise was lost."

Strictly all roses are shrubs, but are included here since they have come to be regarded as an essential part of the perennial garden. The many thousands of cultivars, with more being added to them every year, come from a genus of over 100 (or according to some botanists about 250) species widely distributed throughout the temperate and sub-tropical regions of the northern hemisphere. The genus is a vast and complex subject with many obscurities, and only a few of the neglected but garden worthy species are listed here.

R. brunonii (often misnamed *R. moschata*; Musk rose; Himalayas; white), *R. californica* (Western United States; deep pink), *R. damascena* (Damask rose; Europe; pink to crimson), *R. gallica* and varieties (including Apothecary's rose, Red rose of Lancaster; Europe; variable pink), *R. gigantea* (one parent of the tea and hybrid tea roses; Burma and China; creamy white), *R. primula* (Turkestan; primrose yellow), *R. virginiana* (Eastern United States; pink), *R. webbiana* (Himalayas; blush pink), *R. xanthina* (China; yellow), etc. Most of these species and many others have varieties and cultivars; see specialist lists.

ROSCOEA

FAMILY: *Zingiberaceae*

Named after William Roscoe (1753–1831), founder of the Liverpool Botanic Garden, England, 15 species of slightly tender perennials from Asia. The few in cultivation are attractive subjects for the sheltered rock garden, but *R. purpurea* may be planted in the front border.

R. alpina (Himalayas; pink), *R. cautleoides* (China; yellow), *R. humeana* (China; lilac pink to purple), *R. purpurea* (Sikkim; purple) and variety *procera* (white and mauve).

ROSEMARINUS (Rosemary)

FAMILY: *Labiatae*

From the Latin *ros*, dew or spray, and *marinus*, the sea; in reference to the seaside habitat of this genus from Southern Europe. At one time it was considered to consist of only one species but modern botanists now list four or more. *R. officinalis*, common rosemary, was probably introduced to Britain in Roman times; in the Middle Ages it was considered to be a cure for "all manner of evils of the body" and it became a symbol of remembrance—Ophelia mentions it in her sad speech in Hamlet—since the ancients considered it to have a stimulating effect on the mind. *R. officinalis* and most of its varieties (or separate species) are tall shrubs, but one is an attractive and aromatic plant for paving and the rock garden.

R. lavandulaceus (syn. *R. officinalis* variety *prostratus;* blue).

RUDBECKIA   (Cone flower)

FAMILY: *Compositae*

Commemorating Olaf Rudbeck, a Swedish professor of botany and associate of Linnaeus in the first half of the eighteenth century. A genus of about 25 species, mostly perennials but some annuals; all from North America and closely related to *Echinacea.* Useful for the large border.

Perennials: *R. fulgida* (syn. *R. deamii;* yellow with purple black center) with variety *speciosa* (syns. *R. speciosa, R. newmannii*) and cultivars 'Goldsturm,' 'Goldquelle,' 'Herbstsonne,' etc., *R. maxima* (yellow; black center), etc.

Annuals, treated as half hardy where necessary: *R. bicolor* (yellow with purple) with variety *superba* and cultivars 'Golden Flame,' 'Kelvedon Star,' etc., *R. hirta* (strictly a biennial; Black eyed Susan; yellow and dark brown), *R. tetra* (Gloriosa daisy; yellow to mahogany red), etc.

SALPIGLOSSIS   (Painted tongue)

FAMILY: *Solanaceae*

From the Greek *salpin,* a tube, and *glossa,* a tongue; referring to the tongue-like style in the corolla tube. A genus of 18 species of annuals, biennials and perennials of which only one—of Chilean origin and usually treated as a half hardy annual—is seen in gardens. This is so handsome in its velvety coloring that it may be a matter for regret that some of the other species are not brought into cultivation.

S. sinuata, and nurserymen's strains, 'Grandiflora,' 'Splash,' 'Triumph,' etc. (shades of yellow, pink, crimson, purple, etc.)

SALVIA

FAMILY: *Labiatae*

From the Latin *salveo*, meaning save or heal, used by Pliny in reference to the medicinal properties of some species. A large genus of over 700 species of annual, biennial and perennial plants and shrubs widely distributed in the temperate and warmer zones, which includes the culinary plant *S. officinalis*, common sage, and some of our most useful border and bedding plants. There are 20 or so species in cultivation, some for the greenhouse; the following are hardy or, in the case of those grown as annuals, usually treated as half hardy. A selection for the border and bedding.

Perennials or sub-shrubs: *S. argentea* (Mediterranean regions; silvery leaves, white flowers), *S. azurea* (North America; deep blue), *S. officinalis* (Common sage; Europe) in varieties *purpurascens* (purplish foliage) and *aurea* (golden foliage), *S. pratensis* (Europe; blue) with variety *rosea* (pink), etc.

Biennial: *S. sclarea* (Clary; Europe; flowers mauve, bracts white to pink and purple).

194   Annuals or grown as such: *S. farinacea* (Texas;

violet blue), *S. horminium* (Europe; lilac to purple) and cultivars 'Oxford Blue,' 'Pink Sundae,' etc., *S. splendens* (Scarlet sage; Brazil; scarlet) and cultivars 'Fireball,' 'Gipsy Rose,' 'Purple Blaze,' 'Salmon Pygmy,' etc.

SANGUINARIA   (Bloodroot)

FAMILY: *Papaveraceae*

From the Latin *sanguis*, blood; referring to the red sap of this plant. A genus of only one species; a perennial from Eastern North America, and a beautiful but somewhat difficult subject for moist, light shade.

S. canadensis (white) and variety *flore-pleno* (double).

SANGUISORBA

FAMILY: *Rosaceae*

From the Latin *sanguis*, blood, and *sorbere*, to soak up, from the reputed power of these plants to stop bleeding. Three species of tall perennials which do well grown near water.

S. canadensis (North America; creamy white), *S. obtusa* (Japan; rosy purple) and variety *alba* (white), *S. officinalis* (Burnet; Europe; dull crimson).

SANVITALIA

FAMILY: *Compositae*

Named in honor of the Sanvitali family of Parma, Italy. A genus of seven species from North America and Mexico of which only one is in cultivation. A low growing and spreading annual, useful for ground cover and best treated as half hardy in most areas.

S. procumbens (Mexico; yellow with a dark brown disk) and variety *flore-pleno* (double).

SAPONARIA

FAMILY: *Caryophyllaceae*

From the Latin *sapo,* soap; since the crushed leaves of *S. officinalis* produce a lather with water and were at one time used as a substitute for soap. Some 30 species of perennials and annuals mostly from the Mediterranean regions. Those in cultivation are useful for the rock garden and border, but *S. ocymoides* may seed itself too freely and *S. officinalis* can become invasive.

Perennials for the border: *S. officinalis* (Bouncing Bet, Soapwort; Europe to Japan; pink), and varieties *alba-plena* (double white) and *rosea-plena* (semi-double pink).

For the rock garden: *S. caespitosa* (Pyrenees; pink), *S. ocymoides* (Europe and Caucasus; pink).

Annuals: *S. calabrica* (Italy, Greece; pink), *S. vaccaria* (Europe; pink).

SARRACENIA   (Pitcher plant)

FAMILY: *Sarraceniaceae*

Named after Dr. M. S. Sarrazin of Quebec, who began sending plants to Europe in the seventeenth century. Ten species of evergreen perennials from the Eastern and South Eastern United States, interesting in that they are carnivorous and trap and digest insects. They are usually treated as greenhouse plants, but can be tried outdoors in particularly favored localities. *S. purpurea* is the hardiest

and is said to be naturalised now in bogs in Ros-
common and West Meath in Eire.

S. drummondii (pitchers green and purple, purple
flowers), *S. flava* (pitchers yellowish green, flowers
yellow), *S. minor* pitchers green to purple, flowers
yellow), *S. psittacina* (pitchers veined purple, purple
flowers), *S. purpurea* (pitchers green to purple,
flowers purple).

SAXIFRAGA   (Rockfoil)

FAMILY: *Saxifragaceae*

From the Latin *saxum,* stone, and *frango,* to break;
alluding either to the supposed ability of the roots
to break down rocks, or to the one time medicinal
use of infusions from some of the plants for the
treatment of stones in the bladder. A genus of
about 370 species of mainly dwarf perennials from
the mountain regions of both north and south
temperate zones. It is subdivided into 16 sections
and its complications of nomenclature and innumerable varieties and hybrids are again a subject
for the specialist; species range from the ubiqui-
tous and easily grown to the rare and extremely
difficult, but it is safe to say that there are few rock
gardens, however small, which will not be found
to contain at least one saxifrage. The following is
a short representative selection of the most reliably
named.

S. aizoides (Europe; yellow), *S. burseriana* (Eastern
Alps; white), *S.* x *elizabethae* (hybrid; yellow), *S.
granulata* (Fair maids of France, Meadow saxifrage;
Europe; white), *S. grisebacchii* (Greece; crimson), *S.*   *197*

longifolia (Pyrenees; white), *S. moschata* (Europe; white, pink or yellow), *S. oppositifolia* (Britain and Europe; pink magenta), *S. stolonifera* (syn. *S. sarmentosa;* Mother of thousands; Asia; white spotted pink), *S. umbrosa* (London pride; Britain and Europe; pinkish), etc.

All of the above have many varieties and hybrids.

SCABIOSA

FAMILY: *Dipsaceae*

From the Latin *scabies*, itch, for which some of these plants were used as a cure; or from *scabiosus*, rough, describing the grey felting on the leaves of certain species. About 100 species of annual and perennial plants mainly from the Mediterranean regions; those in cultivation are well known border and florists' subjects, but there are two lesser known species which are good for the rock garden.

Perennials for the border: *S. caucasica* (Caucasus; bluish mauve) with variety *alba* (white) and Cultivars 'Clive Greaves,' 'Miss Willmott,' etc., *S. columbaria* (Europe; lilac), *S. ochroleuca* (South-eastern Europe; pale yellow), *S. succisa* (syn. *Succisa pratensis;* Devil's bit scabious; Britain; white to blue purple), etc.

For the rock garden: *S. graminifolia* (Southern Europe; pale mauve to rose), *S. ochroleuca* variety *webbiana* (creamy white), etc.

Annual: *S. atro-purpurea* (Sweet scabious, Mourning bride, Mournful widow, Pincushion flower;

South-western Europe; deep crimson to purple) and many cultivars, 'Azure Fairy,' 'Blue Moon,' 'Black Prince,' 'Fire King,' 'Parma Violet,' etc.

SCHIZANTHUS   (Butterfly flower;)
               (Poor man's orchid)

FAMILY: *Solanaceae*

From the Greek *schizo,* to split, and *anthos,* flower; referring to the deeply cut corolla, or flower petals. A genus of 15 species of annual plants from Chile, generally grown for greenhouse decoration but also treated as half hardy annuals and used for summer bedding.

S. grahamii (lilac, rose and yellow), *S. pinnatus* (violet and yellow), *S. retusus* (rose and orange) and many strains evolved from their hybrids; 'Danbury Park Strain,' 'Dr. Badger's Hybrids,' 'Dwarf Bouquet,' etc., in a wide color range. See seedsmen's lists.

SCHIZOSTYLIS   (Kaffir lily)

FAMILY: *Iridaceae*

From the Greek *schizo,* to split, and *stylos,* a style; referring to the deeply divided style. Two species of rhizomatous plants from South Africa, of which only one is in cultivation. This is not quite hardy but valuable in the milder winter areas where it can be grown, since it blooms late in the year when other flowers are becoming scarce.

S. coccinea (crimson) and varieties 'Mrs. Hegarty' (pink) and 'Viscountess Byng' (paler crimson).   *199*

SCILLA

FAMILY: *Liliaceae*

An ancient Greek or Latin name used for *Urginea maritima*, the "Sea squill," or from the Latin *squilla*, the squill. A genus of 80 species of small bulbous plants from Europe, Asia and Africa, with a few from tropical Africa. The English and Spanish bluebells were at one time known as scillas but have now been renamed *Endymion;* the scillas are also closely related to *Chionodoxa*. They are invaluable for naturalising, planting in drifts in front of the border or massing in the rock garden. The following are the species usually offered.

S. amoena (Byzantine squill, Star hyacinth; Central Europe; indigo), *S. autumnalis* (Europe and North Africa; rose-lilac), *S. bifolia* (Southern Europe; blue, sometimes purplish or white), *S. pratensis* (Yugoslavia; bluish lilac), *S. sibirica* (Siberian squill; Siberia; brilliant blue), *S. verna* (Europe; blue-mauve), etc.

SCUTELLARIA

FAMILY: *Labiatae*

From the Latin *scutella*, a small shield or dish, referring to the form of the calyx. A widely distributed genus of 300 species of which those in cultivation include some warm house plants and a few uncommon perennials listed here for the rock garden.

S. alpina (South-east Europe; purple), *S. baicalensis* (Russia; blue), *S. indica* (a misnomer since the species comes from China; blue) and variety *japon-ica* (Japan; violet purple), etc.

SEDUM

FAMILY: *Crassulaceae*

From the Latin *sedo,* to sit; referring to the manner in which some species attach themselves to stones and walls. Some 600 annual, biennial or perennial succulent plants; mostly natives of the temperate or colder regions of the Northern hemisphere, with exceptions from central Africa or Peru. A few are regarded as choice plants but most have only the advantage that they will flourish in places where little else will grow. The native British *S. acre,* "Stonecrop," can become an invasive weed; on the other hand *S. spectabile* is extremely attractive to butterflies in the late summer and the annual *S. caeruleum* is a charming little plant which will sometimes seed and colonise itself harmlessly in the rock garden. The following is a selection of the best from the large number of hardy species

Perennial: *S. aizoon* (Asia; yellow), *S. divergens* (North America; purplish leaves, yellow flowers), *S. floriferum* (China; yellow) and cultivar 'Weihenstephener Gold,' *S. hispanicum* (Europe and Persia; white), *S. oreganum* (Western North America; golden yellow), *S. spectabile* (China; pink) and cultivar 'Brilliant' (darker pink), *S. spurium* (Caucasus; white to crimson) and cultivar 'Schorbusser Blut,' etc.

Annual: *S. caeruleum* (Southern Europe; blue).

SEMPERVIVUM   (Houseleek)

FAMILY: *Crassulaceae*

From the Latin *semper*, always, and *vivo*, I live; an allusion to the tenacity of these plants, or the fact that again they will grow almost anywhere and remain, most of them, looking imperturbably the same winter and summer. There are 25 species of hardy succulent perennials mostly from the mountains of Europe; but many of them together with their numerous hybrids and varieties, defy accurate identification. The following however are distinct species.

S. arachnoideum (Cobweb houseleek; Pyrenees and Alps; pink) and hybrids, *S. arenarium* (Tyrol; pale yellow), *S. ciliosum* (Bulgaria; greenish yellow), *S. grandiflorum* (Switzerland; yellow with violet markings), *S. ingwersenii* (Caucasus; pink and white), *S. tectorum* (St. Patrick's cabbage; European Alps; leaves tipped purple, flowers pinkish red) variable in its hybrids and varieties, etc.

SENECIO

FAMILY: *Compositae*

From the Latin *senex*, an old man; an allusion to the grey-haired seed pappus. The largest and most widespread genus in the plant world, containing between 2,000 and 3,000 species and no doubt ultimately to be split up into several separate genera. It covers the widest possible range of plant types including greenhouse and hardy annuals, perennials, evergreens, climbers and shrubs, one aquatic species, a number of species of tree-like dimensions, while the showy and popular green-house cinerarias are hybrids of *S. cruentus*. The following is a selection of the comparatively few hardy annual and herbaceous perennial species in cultivation; all are for the border.

Perennials: *S. doronicum* (Europe; yellow), *S. incanus* (Europe; yellow), *S. palustris* variety *aurantiacus* (Europe; yellow to orange), *S. pulcher* (Uruguay; red-purple), *S. smithii* (Falkland Islands; white), etc.

Annuals: *S. arenarius* (South Africa; lilac), *S. elegans* (syn. *Jacobaea elegans;* South Africa; variable).

SHORTIA

FAMILY: *Diapensiaceae*

Named after Dr. Charles Wilkins Short of Kentucky, a nineteenth century botanist. A genus of about 8 species of dwarf perennials related to *Galax* and natives of China, Formosa and Japan with one from the United States. The two or three in cultivation are choice and beautiful plants for the lightly shaded rock garden, but unfortunately somewhat difficult.

S. galacifolia (Oconee bells; North Carolina; white tinged with pink), *S. uniflora* (Nippon bells; Japan; shell pink) and variety *grandiflora* (sometimes white).

SIDALCEA

FAMILY: *Malvaceae*

A compound of *Sida* and *Alcea*. The first comes from an ancient Greek name used by Theophrastus for the water lily, the second from *Alcea*, a mallow, since these hardy perennials belong to the same family. There are 25 species all from Western North America, and those in cultivation are excellent plants for the border.

S. candida (Colorado; white), *S. malvaeflora* (California; lilac) with variety *listeri* (pink), *S. spicata* (Western North America; rosy purple), and cultivars 'Brilliant,' 'Interlaken,' 'Rev. Page Roberts,' 'Rose Queen,' etc.

SILENE   (Catchfly)

FAMILY: *Caryophyllaceae*

Probably from the Greek *sialon*, saliva; referring to the gummy exudation on the stems. A large genus of 500 species of annuals, biennials and perennials from the Northern hemisphere and South Africa. Few of them are particularly distinguished and some are simply weeds, but the following are good garden subjects; annual and biennial species for the border or edging, and perennials mostly for the rock garden.

Perennials: *S. acaulis* (Moss campion; widespread in the north temperate zones; pink), *S. elizabethae* (Tyrol; crimson), *S. laciniata* (syn. *Melandrium laciniatum*; North America; scarlet), *S. schafta* (Caucasus; pink), *S. virginica* (North America; crimson), etc.

Biennials: *S. compacta* (Russia and Asia Minor; pink), *S. rupestris* (Europe and Siberia; white to pink).

Annuals: *S. armeria* (Europe; pink), *S. pendula* (Europe; white to rose) with varieties and cultivars, 'Peach Blossom,' 'Triumph,' 'Special Dwarf Mixture,' etc.

SILPHIUM   (Rosinwood)

FAMILY: *Compositae*

An ancient Greek name referring to the resinous juice of these plants. A genus of 15 species of perennials from North America of which only two are likely to be seen in cultivation.

S. laciniatum (Compass plant, the leaves said to face north and south; yellow), *S. perfoliatum* (Cup plant, the leaves being joined together to make a cup round the stem; yellow).

SISYRINCHIUM

FAMILY: *Iridaceae*

An old Greek name probably first applied to some other plant. A genus of 100 species of rush like annuals or perennials related to *Iris* and mostly from the Americas. They are somewhat insignificant, but the few perennial species in cultivation are useful for corners of the rock garden or planted between paving stones. Those listed here are hardy.

S. angustifolium (Blue eyed grass; North America; blue), *S. bellum* (North America; violet), *S. douglasii* (North America; violet blue), *S. filifolium* (Falkland Islands; white), *S. striatum* (Argentine and Chile; yellow), etc.

SMILACINA

FAMILY: *Liliaceae*

The diminutive of *Smilax* (a genus of greenhouse plants which it resembles), literally "a little smilax," which is itself a Greek name of obscure meaning. A genus of 25 species of hardy rhizomatous perennials from America and Asia. Only three are in   *205* cultivation, and they are best in moist positions in the shaded wild garden.

S. oleracea (Sikkim; white and pale pink, rose-purple berries), *S. racemosa* (False spikenard; North America; white, red berries), *S. stellata* (North America; white).

SOLANUM   is a large and diverse genus of some 1,700 species, including the potato, aubergine and other edible fruit plants, together with some which are poisonous. The best of the decorative species are either greenhouse subjects or large shrubs.

SOLDANELLA   (Blue moonwort)

FAMILY: *Primulaceae*

From the Italian *soldo*, a small coin; referring to the round leaves of some of the plants in this genus. A few species of small hardy perennials from central and southern European mountains. They are among the most delicate and beautiful of spring flowering plants for the rock garden.

S. alpina (blue marked with crimson), *S. minima* (blue with violet), *S. montana* (violet), *S. pusilla* (blue or violet blue), etc.

SOLIDAGO   (Golden rod)

FAMILY: *Compositae*

From the Latin *solido*, to make whole, or heal; referring to the medicinal value of some of these plants. A genus of over 100 species of perennials mostly from North America, with a few from Europe. The taller are useful for the large border, 206 although *S. canadensis* can become invasive, and there are some varieties suitable for the rock garden. All have flowers in shades of yellow.

For the border: *S. canadensis* (North America), *S. graminifolia* (North America), *S. odora* (North America; aromatic foliage), *S. virgaurea* (Europe), etc. Cultivars include 'Golden Wings,' 'Lemore,' 'Lesden,' 'Peter Pan,' etc.

For the rock garden: *S. virgaurea* varieties *minuta, alpestris* and *cambrica.*

SPARAXIS   (Harlequin flower)

FAMILY: *Iridaceae*

From the Greek *sparasso,* to tear; in reference to the torn appearance of the spathes. A few species of cormous perennials from South Africa which are strictly cool greenhouse subjects. Those offered by bulbsmen under such names as 'Tricolor Choice Mixture,' 'Fire King,' 'Sulphur Queen,' etc., are probably bigeneric hybrids between *Sparaxis grandiflora* or *S. tricolor* and *Streptanthera cuprea.* They appear to be slightly hardier than the species and are suitable for sheltered positions in mild winter areas. Their common name "Harlequin flower" aptly describes their gay and often patterned coloring.

SPECULARIA   (Venus's looking glass)

FAMILY: *Campanulaceae*

From the Latin *speculum,* a mirror; since "Venus's looking glass" is the old common name of one of the species. A small genus from the Northern hemisphere once known, and still so regarded by many botanists, as *Legousia.* Only one is now commonly   207 grown, an attractive little annual for the front border.

S. speculum (syn. *Campanula speculum;* Europe; blue) with variety *alba* (white).

SPIRAEA   is a genus of shrubs.

STACHYS

Family: *Labiatae*

From the Greek *stachus,* a point; referring to the shape of the flower head. A diverse genus of 300 species of annuals, perennials, sub-shrubs and shrubs widely distributed throughout the world. Most of them are weeds, but the few perennials in cultivation are useful for the rock garden or border.

For the border: *S. coccinea* (Central America; scarlet), *S. macrantha* (syns. *S. grandiflora, Betonica macrantha;* Betony; Caucasus; violet), *S. officinalis* (syn. *S. betonica, Betonica officinalis;* Bishop's wort, Wood betony; Europe and Asia Minor; purple), etc.

For the rock garden: *S. corsica* (Corsica and Sardinia; cream and pink), *S. lanata* (Lamb's ear; Caucasus to Persia; grey woolly foliage, flowers purple), *S. lavandulifolia* (Armenia; purplish rose), etc.

STATICE   See Limonium

STERNBERGIA

Family: *Amaryllidaceae*

Named after Count Kaspar Moritz von Sternberg, a botanist of Prague in the late eighteenth and early nineteenth centuries. A genus of eight species of bulbous plants from the Mediterranean regions, Central Europe and Asia Minor. They resemble both *Crocus* and *Colchicum* in form although both belong to entirely different families. Three of the four species generally in cultivation and listed here flower in the autumn: the other, *S. fischeriana,* is a spring-flowering type. All are bright yellow.

S. colchiciflora (Hungary), *S. fischeriana* (Mediterranean regions), *S. lutea* (Asia Minor), *S. clusiana* (syn. *S. macrantha;* Southern Europe).

STOKESIA

FAMILY: *Compositae*

Named after Dr. Jonathan Stokes, a doctor of medicine and botanist of the eighteenth and nineteenth centuries. There is only one species, a late flowering perennial from the Southeastern United States, and an attractive plant for the front border in mild winter areas.

S. laevis (syn. *S. cyanea;* blue) with varieties *alba, elegans, praecox,* and *rosea* (variable white to pink and mauve).

STREPTANTHERA

FAMILY: *Iridaceae*

From the Greek *streptos,* twisted, and *anthera,* anther; the flowers having curiously shaped anthers. A genus of only two species of cormous plants from South Africa, and closely related to *Sparaxis* and *Ixia* with which they will hybridise.

S. cuprea (coppery, purple center) with variety

coccinea (orange and black), *S. elegans* (white, purple center).

See Sparaxis.

STREPTOCARPUS   is a genus of tropical plants, of which most cultivated forms are hybrids and usually treated as greenhouse annuals.

STYLOPHORUM
FAMILY: *Papaveraceae*

From the Greek *stylos*, style, and *phero*, bearing; referring to the style being retained on the seed capsule after the flower has withered. Three species of perennials from Asia and Eastern North America and useful plants for the partially shaded border.

S. diphyllum (Celandine poppy; Eastern North America; yellow), *S. lasiocarpum* (Eastern China; yellow to reddish), *S. sutchuenense* (Western China; yellow).

SYMPHYANDRA
FAMILY: *Campanulaceae*

From the Greek *symphio*, together or uniting, and *andros*, anther; in reference to the formation of the anthers. A genus of ten species of perennials from the Eastern Mediterranean to the Caucasus and Central Asia. They are similar to the campanulas but separated from them botanically by the different form of the anthers. Most of those in cultivation are for the border but *S. wanneri* is a good plant for the rock garden.

S. asiatica (Asia; violet), *S. cretica* (Mediterranean regions; lilac), *S. hofmannii* (Balkans; white), *S. pendula* (Caucasus; yellow), *S. wanneri* (South-eastern Europe; blue).

SYMPHYTUM

FAMILY: *Boraginaceae*

From the Greek *symphio*, to unite; a reference to the one time use of these plants for healing wounds. About 25 species of perennials generally distributed from Europe to the Caucasus. They are large, spreading plants, apt to become invasive and best suited for the wild garden.

S. asperum (Prickly comfrey; rose, turning blue), *S. caucasicum* (blue), *S. grandiflorum* (yellowish white), *S. officinale* (Common comfrey) and varieties *argenteum* (leaves variegated white), *coccineum* (crimson), *ochroleucum* (pale yellow to white), *purpureum* (purple), *S. peregrinum* (rose changing to blue), etc.

SYNTHYRIS

FAMILY: *Scrophulariaceae*

From the Greek *syn*, united, and *thyris*, a door; a reference to the closed valves of the seed capsule. A genus of 15 low-growing perennials from the mountains of Western North America. They are small attractive plants for the sheltered rock garden where the soil does not dry out in summer, and of the several species said to be in cultivation the one usually offered by nurserymen is probably the best.

S. reniformis (violet blue).

TAGETES

FAMILY: *Compositae*

From the Latin, and said to be after Tages, the legendary founder of ancient Etruscan augural law; a grandson of Jupiter, who is supposed to have sprung from the ploughed earth in the form of a boy and to have taught the Etruscans the meaning of lightning, winds, eclipses, and other signs. A genus of 50 species of annuals and perennials mostly from Mexico and the Southern United States, and not to be confused with the English or "Pot marigold" *Calendula officinalis.* Similarly the descriptions "African" and "French" are misnomers for these Mexican and Southern American plants, and may have been applied as a result of them coming into England from France where, according to Gerard, they were introduced by Charles V of the Holy Roman Empire—who probably received them from Hernándo Cortéz on his return to Spain in 1528 after the conquest of Mexico. Today, however, with a few exceptions the species have been virtually lost in the wide range of hybrids, cultivars and strains which, treated generally as half hardy annuals, will succeed almost anywhere in the garden. All the following are naturally annual.

T. erecta (African marigold; lemon yellow to orange) with strains and cultivars, 'All Double Orange,' 'Carnation Flowered Alaska,' 'First Lady,' 'Dixie Sunshine,' 'Guinea Gold,' 'Hawaii,' etc., *T. lucida* (Mexican marigold; yellow), *T. patula* (French marigold; brownish yellow) with strains and cul-

tivars in a wide range of orange, red and mahogany shades; 'Harmony,' 'Legion of Honor,'
'Naughty Marietta,' 'Red Brocade,' etc., *T. tenuifolia*
(syn. *T. signata*) variety *pumila*, with strains and
cultivars 'Gnome,' 'Golden Gem Selected,' 'Lulu,'
'Irish Lass,' etc. See seedsmen's lists.

TANACETUM is a genus which contains the
medicinal herb "Tansy," *Tanacetum vulgare*, but
neither this nor other species is of any garden
value.

TECOPHILAEA   is a genus of two or three species
of small cormous plants resembling crocuses, from
Chile. They are rare, difficult and beautiful in
about equal proportions. High class bulbsmen
sometimes tempt you with the corms, but when
they can be grown at all they need the shelter of
a cool greenhouse.

TEUCRIUM   (Germander)

FAMILY: *Labiatae*

Named after Teucer, a legendary king in the region of Troy, and yet another who is said to have
used some of these plants in medicine. About 300
species of perennials and shrubs widely distributed in the warmer temperate parts of the world
especially the Mediterranean. Few of them are of
any distinction, but some of the perennials and
dwarf shrubs are suitable for the large rock garden.

   T. chamaedrys (Europe; rosy red), *T. pyrenaicum*
(Pyrenees; lilac and white), *T. scordium* (Mediterranean area; yellow), etc.   213

THALICTRUM   (Meadow rue)

FAMILY: *Ranunculaceae*

From the Greek *thaliktron*, the name used to describe a plant with divided leaves. A genus of 150 perennials, mainly from north temperate regions but also represented in tropical South America, tropical Africa and South Africa. Those in cultivation are mostly for the border, the dwarf species being best in the alpine or cold greenhouse.

T. aquilegifolium (Europe and Northern Asia; pale purple), *T. dipterocarpum* (Western China; deep lavender) with variety *alba* (white) and cultivar 'Hewitt's Double' (violet mauve), *T. glaucum* (syn. *T. speciosissimum*; Europe and North Africa; yellow), etc.

THERMOPSIS

FAMILY: *Leguminosae*

From the Greek *thermos*, a lupin, and *opsis*, like; referring to this plant's close resemblance to the lupin. A genus of some 30 species of perennials from North America and Asia. Those in cultivation are all easy and attractive plants for the border.

T. barbata (Himalayas; purple), *T. caroliniana* (Eastern United States; yellow), *T. lupinoides* (Siberia; yellow), *T. mollis* (Eastern United States; yellow), *T. montana* (Western North America; yellow).

THLASPI

FAMILY: *Cruciferae*

From *thlaspis*, the ancient Greek name for a
214   cress. About 60 species of small annual, biennial

and perennial cresses widely distributed in the north temperate regions. Most of them are worthless, but there are one or two perennials from the European Alps which are attractive in the rock garden.

T. alpinum (white), *T. rotundifolium* (rosy lilac).

THYMUS   (Thyme)

FAMILY: *Labiatae*

Possibly from the Greek *thymos*, thyme; itself derived from *thuo*, to perfume. Between 300 and 400 species of dwarf aromatic shrubs and subshrubs distributed widely throughout Europe and Asia. The culinary species are *T. citriodorus* and *T. vulgaris* with variety *aureus*, but there are many other flowering aromatic plants for the rock garden and pavements of which the following is a selection.

T. caespititius (Spain, the Azores and Madeira; mauve), *T. comosus* (Eastern Europe; pink), *T. doerfleri* (Yugoslavia; rose purple), *T. hirsutus* (the Balkans; lilac to pink), *T. serpyllum* (Europe; purple), with varieties *albus, coccineus,* and *roseus,* and cultivars 'Annie Hall," 'Pink Chintz,' etc.

TIARELLA

FAMILY: *Saxifragaceae*

From the Greek *tiara,* a little turban, describing the shape of the seed capsule. A small genus of perennial plants from the Himalayas, North America and Eastern Asia. They are quietly attractive in lightly shaded positions where they can be allowed space to spread. *T. cordifolia* has been crossed with a hybrid *Heuchera* to give the bi-generic hybrid x *Heucherella*.

T. cordifolia (Foam flower; North America; white), *T. polyphylla* (India, China and Japan; white or pink), *T. trifoliata* (Oregon to Vancouver; pale pink), *T. wherryi* (Southern United States; white or pale pink).

TIGRIDIA   (Shell flower; Tiger flower)

FAMILY: *Iridaceae*

From the Greek *tigris,* tiger, and *eidos,* like; de-scribing the strikingly marked flowers. A small genus of bulbous plants from Mexico, Chile and Peru of which only one species is in general culti-vation. Tigridia was probably brought into Europe by a physician whom Philip II of Spain sent to Mexico in the second half of the sixteenth century, who wrote that it grew wild there and was much prized for its beauty and medical virtues, being a 'fridgefacient' in fevers and also a promoter of fecundity in women. It is another plant which produces only one flower at a time, although six or more will appear in succession, and this brief glory moved another much later writer—in the London *Botanical Magazine* for 1831, when it was illustrated as *Ferraria tigridia*—to exclaim ". . . in splendid beauty it appears to us to surpass every competitor; we lament that this too affords our fair countrywomen another lesson: how extremely fugacious is the loveliness of form, born to display its glory but for a few hours, it literally melts away." But, however fugacious, groups of a dozen

216

or more set together are among the showiest of plants for the sunny border.

T. pavonia (shades of yellow to orange and crimson to purple, centrally blotched with maroon).

TITHONIA   (Mexican sunflower)

FAMILY: *Compositae*

From Tithonus, son of Laomedon and brother of Priam of Troy; he was loved by Eos, goddess of the dawn, who begged Zeus to make him immortal but forgot to ask also for eternal youth to go with the gift. A small genus from South America, Central America and the West Indies of which only two species are in cultivation. They are large and showy late flowering plants for the back of the border, best treated as half hardy annuals in all but the most favorable climates.

T. diversifolia (yellow), *T. rotundifolia* (syns. *T. speciosa, T. tagetifolia;* orange red) and cultivars 'Torch,' 'Fireball,' etc.

TRADESCANTIA

FAMILY: *Commelinaceae*

Commemorating John Tradescant, who died in 1637; one of the greatest and most adventurous of the early plant collectors, who became gardener to King Charles I, founded a botanic garden at Lambeth in London and brought many new plants into England, including the gladiolus, apricot and others; see *Polyanthus.* The genus which bears his name consists of about 60 species of hardy and greenhouse perennial plants from North America and tropical South America, of which the only   *217* hardy perennial generally seen in gardens was imported into England from Virginia by Tradescant. The modern forms of this popular and accommodating border plant are now said by some botanists to belong to a hybrid group which should be known as *T* x *andersoniana.*

T. virginiana (or *T* x *andersoniana;* Devil in the pulpit, Flower of a day, Moses in the bulrushes, Spiderwort; Eastern United States; violet-blue) with varieties *alba* (white), *caerulea* (bright blue), *rosea* (pink), *rubra* (ruby red) and cultivars 'Iris Pritchard,' 'J. C. Weguelin,' 'Osprey,' etc.

TRICYRTIS   (Toad lily)

FAMILY: *Liliaceae*

From the Greek *treis,* three, and *kyrtos,* convex; in reference to the three outer petals which are convex at the base. A genus of ten species of perennials from Japan, Formosa, and the Himalayas. They are curious rather than beautiful on account of their oddly shaped flowers and bizarre coloring, but interesting for a warm, partly shaded position in woodland or the rock garden.

T. bakeri (origin uncertain but possibly Japan; yellow patterned with purple), *T. formosana* (Formosa; lilac spotted purple) and variety *stolonifera,* *T. hirta* (Japan; white, spotted purple, sometimes tinged with pink or green), *T. latifolia* (Japan; cream spotted purple), *T. macrantha* (Japan; yellow spotted purple), *T. maculata* (Himalayas; greenish white, 218   spotted purple).

TRIFOLIUM   (Clover)

FAMILY: *Leguminosae*

From the Latin *tres*, three, and *folium*, leaf, referring to the trifoliate leaves. A genus of some 300 annual and perennial species widely distributed throughout the temperate and sub-tropical Northern hemisphere. Most of the clovers are weeds or little better, but a few of the more uncommon species may find a place in the rock garden.

T. alpinum (European Alps; pink to mauve), *T. repens* variety *purpurascens* (Europe; bronze purple foliage), *T. uniflorum* (Syria; bluish purple), etc.

TRILLIUM

FAMILY: *Liliaceae*

From the Latin *triplum*, triple; referring to the three-part leaves and flowers. A genus of 30 species or more of hardy perennials mostly from North America, but with a few of less interest from Japan and the Himalayas. Beautiful plants for the shaded woodland garden but difficult to establish. *T. grandiflorum* is probably the easiest and best.

T. grandiflorum (Wake robin; Eastern North America; white, becoming pink), *T. sessile* (Toad trillium; North America; crimson purple) with varieties *californicum* (white to purplish pink), *luteum* (yellow) and *wrayi* (variable purple), *T. undulatum* (Painted trillium, Painted wood lily; Eastern North America; white marked with crimson or purple), etc.

TRITONIA is a genus of South African cormous plants allied to *Ixia*, *Sparaxis*, *Streptanthera* and *Freesia* and like most of these they are only safe in the cool greenhouse in most areas.

TROLLIUS (Globe flower;)
(Giant buttercup)

FAMILY: *Ranunculaceae*

From the German vernacular name *trollblume*, or the Latin *trulleus*, a basin; describing the shape of the flower. A genus of 25 hardy perennials from the north temperate and arctic regions; easy and attractive plants which are at their best beside a pool or stream.

T. europaeus (Europe; lemon yellow) and varieties 'First Lancers,' 'Goldquelle,' 'Helios,' 'Orange Princess,' etc., *T. ledebouri* (syn. *T. chinensis*; Northern China; deeper yellow) and cultivars including 'Golden Queen,' 'Imperial Orange,' etc., *T. yunnanensis* (Western China; yellow), etc.

For the lightly shaded rock garden: *T. pumilus* (Himalayas; buttercup yellow).

TROPAEOLUM

FAMILY: *Tropaeolaceae*

From the Latin *tropaeum*, a trophy; supposedly in allusion to the likeness of the flowers and leaves to helmets and shields as they were displayed at Roman triumphs. A diverse genus of 90 or so species of annual and perennial climbing plants from Mexico and temperate South America which, besides several exotic greenhouse subjects, includes the common garden nasturtium and a somewhat perverse perennial known as "Flame flower." Some of the species which are naturally perennial are treated as annuals in cultivation.

Perennials: *T. speciosum* (Flame flower; Chile; brilliant scarlet), *T. tuberosum* (Peru; orange scarlet), etc.

Hardy or half hardy annuals or treated as such: *T. majus* (Nasturtium; Peru; yellow and orange to crimson) with numerous cultivars 'Golden Gleam,' 'Indian Chief,' 'Orange Gleam,' 'Scarlet Gleam,' etc., and variety *nanum* with cultivars 'Empress of India,' 'Jewel Mixed,' 'Rosy Morn,' etc., *T. peregrinum* (syn. *T. canariense;* Canary creeper; Peru; bright yellow), etc. See seedsmen's lists.

TULIPA   (Tulip)

FAMILY: *Liliaceae*

Either from the Persian *thoulypen* or from the Turkish *tulipam,* both meaning turban and referring to the shape of the flower. A genus of, so far, about 100 species of bulbous plants from the temperate regions of Europe and Asia. They range from a few inches in height to nearly three feet, and from some which produce several tiny flowers on one stem to those which display a single bloom a foot or more in diameter when opened out. The tulip was unknown in Europe until a diplomatic mission returning from Turkey brought back the first bulbs in the middle sixteenth century, and the development of the tulip as we know it today almost certainly dates from 1571 when Carolus Clusius, Professor of Botany at Leyden in Holland, began   *221*

to grow the bulbs in his own small garden; probably *T. clusiana*, the "Lady Tulip." By 1634 tulips
had become a mania in Europe—probably one of
the most extraordinary outbursts of speculation in
history—and prices of £250 to more than £500
were paid for single bulbs in the hope of realising
a profit on their multiplication; there was a house
in Haarlem long known as the "Tulip House" because it was said to have been bought by one such
bulb. Many people were bankrupted when the
crash came a few years later, but by then tulip
growing and breeding was settling down as a
stable Dutch industry, and today it produces upwards of 4,000,000,000 actual bulbs yearly. There
are over 3,000 cultivars now recognised, 800 of
them in commercial production in 23 main groups
or classes. These are best seen in bulbsmen's catalogs; those listed here are only a few of the less
known species which are becoming increasingly
popular as choice plants for many different selected positions in the garden.

   T. biflora (Caucasus; white and yellow), *T. clusi-
ana* (Lady tulip; Persia; white, outer petals pink),
T. eichleri (Baku; scarlet), *T. hageri* (Greece; buff red
overlaid green), *T. praestans* (Bokhara; vermilion),
T. pulchella (Turkey; purplish pink), *T. saxatilis*
(Crete; rosy lilac), *T. tarda* (syn. *T. dasystemon*;
Turkestan; soft yellow), *T. turkestanica* (Turkestan;
white shaded pale green), etc.

   For varieties, groups and cultivars see specialist
bulbsmen's catalogs.

URSINIA

FAMILY: *Compositae*

Named after Johan Ursinus of Regensburg, the seventeenth century author of *Arboretum Biblicum*. A genus of 80 species of annuals, perennials and sub-shrubs from South Africa, with one found in Abyssinia. Only two are generally in cultivation, and these or their cultivars are usually treated as half hardy annuals. They are good for bedding or the front border.

U. anethoides (syn. *Sphenogyne anethoides;* orange yellow purple center) and cultivar 'Aurora' (bright orange and crimson), *U. versicolor* (syn. *U. pulchra;* orange, dark center) and cultivar 'Golden Bedder.'

VALERIANA is a genus of little garden interest. Some species were once grown as medical herbs.

VENIDIO-ARCTOTIS

FAMILY: *Compositae*

A bigeneric hybrid between the genera *Venidium* and *Arctotis,* the flowers in soft shades of orange, pink and buff. See nurserymen's lists.

VENIDIUM

FAMILY: *Compositae*

The origin of the name is obscure; but possibly from the Latin *vena,* a vein, referring to the ribbed seed pods. About 30 species of annuals and perennials from South Africa, and except that the plants are taller they resemble the other South African daisy types. The few species in cultivation are usually treated as half hardy annuals.

223

V. decurrens (yellow, dark center), *V. fastuosum* (Namaqualand daisy; orange, purple zone, black center) and strain 'Dwarf Hybrids' (cream, yellow and orange, etc.).

VERATRUM   (False hellebore)

FAMILY: *Liliaceae*

From the ancient Latin name of hellebore. A genus of 25 species of tall perennials from the north temperate regions, best in the wild garden. Their roots are poisonous.

V. album (White hellebore; Europe and Asia; yellowish white), *V. californicum* (Western North America; white), *V. nigrum* (Southern Europe, Asia; purplish black), *V. viride* (Indian poke; North America; greenish).

VERBASCUM   (Mullein)

FAMILY: *Scrophulariaceae*

Probably a corruption of the Latin *barbascum,* a hairy plant (*barba,* a beard), many species being covered with hair or down. A genus of 300 biennials or short lived perennials from the temperate parts of Europe and Asia. They are tall, handsome plants for sunny positions, though the heavily felted species are curious rather than attractive.

Perennials: *V. nigrum* (syn. *V. vernale;* Europe; yellow and brown), *V. olympicum* (Greece; grey hair, yellow), *V. phoeniceum* (Europe; variable red to purple) with cultivars (pink to lilac), *V. pulverulentum* (Hoary mullein; Europe; white hair, yellow), etc.

224

Biennials or treated as such: *V. bombyciferum* (Asia Minor; thick silvery hair, yellow), *V. chaixii* (Europe; whitish hair, yellow), *V. thapsus* (Aaron's rod, Hag's taper; Europe and Asia; thick white hair, yellow), and cultivars, etc.

VERBENA

FAMILY: *Verbenaceae*

Thought to be from the Latin *verbenae*, the sacred branches of laurel, myrtle or olive carried by heralds and certain priests; or a corruption of the Celtic name *fervain* for *V. officinalis*. Fervain or "Vervain" was the holy herb used in ancient sacred rites; it was also supposed to cure scrofula and the bite of rabid animals, to arrest the diffusion of poison, to avert antipathies and to be a pledge of mutual good faith—hence it was worn as a badge by heralds and ambassadors in ancient times. The genus consists of 250 widely distributed species of annuals and perennials, but is best known by the wide range of florists' verbenas, *V x hybrida*. The perennials in cultivation, nearly all coming from South America, are only hardy in favorable climates and the species listed here are usually treated as half hardy annuals for summer bedding.

V. canadensis (North America; lilac or purple), *V. rigida* (Argentine; claret purple), *V x hybrida* (variable) and cultivars from these and others, 'Compacta Amethyst,' 'Miss Susie Double,' 'Olympia,' 'Royal Bouquet,' etc. See seedsmen's lists.

VERONICA   (Speedwell)

FAMILY: *Scrophulariaceae*

The origin of the name is doubtful, but it is possibly after St. Veronica; the woman who gave Christ a cloth to wipe His face while on the way to Calvary. A genus of about 300 species of annuals, perennials and sub-shrubs mainly from north temperate regions; many of the shrubby species from New Zealand have now been transferred to *Hebe*. Those listed here are useful perennials for the border and rock garden, but *V. filiformis* although beautiful is dangerously invasive and should never be planted where it can do any harm.

For the border: *V. gentianoides* (South-east Europe; blue), with variety *variegata* (variegated leaves; deeper blue), *V. incana* (Europe; blue), with variety *rosea* (pink), *V. longifolia* (Europe; lilac blue) and cultivar 'Foerster's Blue,' *V. spicata* (Europe; bright blue) with many varieties including *alba* (white) and *rosea* (pink), etc.

For the rock garden: *V. cinerea* (Asia Minor; blue or pink), *V. pectinata* (Caucasus; pale blue), with variety *rosea*, *V. prostrata* (Europe; blue) and cultivars 'Mrs. Holt' (pink) and 'Shirley blue,' *V. repens* (Corsica; blue), etc.

For the wild garden: *V. filiformis* (Asia Minor; China blue).

See Nurserymen's catalogs.

VINCA   (Periwinkle)

FAMILY: *Apocynaceae*

The Latin name used by Pliny, probably from 226 *vincio*, to bind; referring to the long tough runners.

A genus of seven species, mostly of trailing perennials widely distributed over Europe and Western Asia. The hardy species are useful for ground cover either in sun or under trees and hedges.

V. major (Britain to the Mediterranean regions and North Africa; blue) with variety *variegata* (leaves margined with white), *V. minor* (Europe; blue) varieties *alba* (white), *atro-purpurea* (deep purple), *aureo-variegata* (leaves blotched with yellow), *azurea flore-pleno* (sky blue, double) and cultivars 'Bowle's Variety,' 'Le Grave,' etc.

VIOLA

FAMILY: *Violaceae*

The old Latin name for violet. A genus of some 500 species, mainly from the north temperate regions which includes *Viola tricolor*, the "Heartsease" of Elizabethan days and one of the parents of the wide range of modern pansies, *V. cornuta* and *V. gracilis* of the strains of garden violas, and *V. odorata* of the violets. Some of the less known species, however, are equally attractive for the rock garden.

For bedding: *V. gracilis* (Viola; Asia Minor and Balkans; deep violet) with varieties *alba* (white), *lutea* (yellow), *major* (deep violet) and hybrid strains usually treated as biennials 'Blue Heaven,' 'Clear Crystals,' 'Perfection Blue,' etc., *V. odorata* (Sweet violet; Europe; blue violet) with varieties including *sulphurea* (pale yellow) and cultivars 'Princess of Wales' (purple), 'Rosina Hybrids' (white to pink and violet), 'The Czar' (blue), etc., *V. tricolor* 227

(Heartsease, Johnny jump up; Europe; purple, yellow and white) and hybrid strains usually treated as biennials, 'Engelmann's Giant,' 'Felix,' 'Masquerade,' 'Pacific Toyland Hybrids,' 'Roggli,' etc. See seedsmen's lists.

Perennials for the rock garden: *V. adunca* (North America; violet or lavender), *V. cornuta* (Pyrenees; violet), and cultivars including the "Violettas" derived mainly from this species (variable yellow, purple and violet), *V. cucullata* (North America; white veined purple), *V. hispida* (Europe; violet), *V. labradorica* (North America; lavender blue), *V. rupestris* (Asia, Europe and North America; blue violet), *V. saxatilis* (Asia Minor and Europe; *violet*), etc. See nurserymen's catalogs.

VISCARIA   See Lychnis

WAHLENBERGIA

FAMILY: *Campanulaceae*

Commemorating Dr. George Wahlenberg of Uppsala, a botanical author in the nineteenth century. About 150 species of widely distributed perennials closely related to *Campanula* and *Edraianthus*. The few sometimes offered by nurserymen are choice plants for sheltered positions on the rock garden.

W. dalmatica (Dalmatia; blue), *W. matthewsii* (New Zealand; white to lilac), *W. saxicola* (syn. *Campanula saxicola*; Tasmania; grey blue), *W. tasmanica* (Tasmania; blue).

WULFENIA

Named after a late eighteenth century Austrian botanist, Franz Xavier Freiherr von Wulfen. A small genus of rhizomatous perennials widely distributed from South-eastern Europe to the Himalayas, of which the species most likely to be seen in cultivation is an interesting plant for the large rock garden.

W. carinthiaca (Eastern European Alps; violet blue).

XANTHISMA   (Star of Texas)

From the Greek *xanthos,* yellow; the flower color. A genus of only one species, a showy annual for the border from the United States.

X. texanum (Texas; yellow).

XERANTHEMUM

From the Greek *xeros,* dry, and *anthos,* a flower. Six species of annuals found from the Mediterranean regions to South-western Asia. There is only one in cultivation; of no particular distinction except that the daisy-like flowers retain their color when dried and are useful for winter decoration.

X. annuum (Mediterranean regions; purple).

YUCCA   (Adam's needle)

The native Peruvian name. A genus of 40 species of greenhouse and hardy evergreens from the   *229*

Southern United States, Mexico and the West Indies. These plants are strictly shrubs but the smaller hardy species in cultivation are included here since they are often grown as spot plants. All have cream flowers.

Y. filamentosa and variety *variegata* (leaves marked with yellow or white), *Y. recurvifolia*, etc.

ZANTEDESCHIA

FAMILY: *Araceae*

Commemorating Francesco Zantedeschi, an Italian physician and botanist in the eighteenth and nineteenth centuries. A small genus of rhizomatous perennials from tropical Africa which was formerly included in *Richardia*, while the white "Arum lily" of florists, *Zantedeschia aetheopica* was placed in *Calla*, as it is still sometimes known. This is the hardiest of the genus in mild winter areas, the yellow and pink flowering species being best suited to pot culture and removal to the garden only during the summer months. It is a handsome plant for the waterside or other moist sunny position.

Z. aethiopica (Arum lily, Calla lily, Lily of the Nile; white) with cultivars 'Crowborough' (hardier than the species) and 'Little Gem,' *Z. albo-maculata* (white, sometimes pale yellow), etc.

ZAUSCHNERIA

FAMILY: *Onagraceae*

Commemorating J. P. J. Zauschner, an eighteenth century professor of natural history at 230 Prague. Four species of sub-shrubby perennials from the Western United States and Mexico, of which the one usually seen in cultivation is an attractive plant for the rock garden in milder winter areas.

Z. *californica* (Californian fuchsia, Humming bird's trumpet; scarlet) and variety *latifolia*.

ZEPHYRANTHES

FAMILY: *Amaryllidaceae*

From the Greek *zephyr*, the west wind, and *anthos*, a flower. A genus of 35 to 40 species of bulbous plants from the warmer parts of America and the West Indies. Most of those in cultivation are for the greenhouse in all but the mildest climates, but the one hardy species is an attractive plant for the front border or rock garden.

Z. *candida* (Flower of the west wind, Zephyr lily; Argentina; white).

ZINNIA

FAMILY: *Compositae*

Named after Johann Gottfried Zinn, an eighteenth century German professor of botany. A genus of 20 species of annuals and perennials from the Southern United States, Brazil and Chile. With the possible exception of *Zinnia linearis* these have now been almost superseded by the numerous and diverse cultivars and strains derived from them and treated as half hardy annuals for bedding.

Z. *elegans* (Mexico; variable yellow to orange and red), Z. *haageana* (Tropical America; scarlet), Z. *linearis* (Mexico; golden yellow and orange),   *231*

Z. *pauciflora* (Mexico; yellow or purple), Z. *tenuiflora*
(syn. Z. *multiflora* (Mexico; scarlet), etc., and cultivars from them 'Chrysanthemum Flowered Hybrids,' 'Button Flowered Mixed,' 'Giant Dahlia
Flowered Strain,' 'Mammoth,' 'Persian Carpet,'
'Thumbelina,' etc. See seedsmen's lists.

A Glossary of Specific Names or Epithets

Although most of the following descriptions are dog Latin, and many of them a Latinised version of some other language, they still follow the basic rule of gender in Latin grammar. A generic title may have a masculine, feminine or neuter ending and the gender of both this and the specific epithet should agree. Taking their gender from that of the genus, descriptive masculine names or epithets usually end in *-us, -is,* or *-er;* feminine in *-a,* or *-ra,* but sometimes in *-is,* or *-ris;* and neuter in *-um, -e,* or *-re;* and thus in many cases the terminals of both generic title and specific epithet will be the same, so producing a pleasant sound and an aid to memory—*Dianthus neglectus, Campanula lactiflora, Geranium grandiflorum,* etc. Due to the differences in gender terminals this is not invariable however, and other exceptions also appear—where old generic names have been adapted as specific epithets, where the epithet itself is a possessive or commemorative name (which see later) or where it is a place name. These usually end in either *-ens* or *-ic* plus the gender of the genus; as in masculine *-ensis* or *-icus,* feminine *-ensis* or *-ica,* and neuter *-ense* or *-icum.*

The following glossary includes the specific names or epithets already used in foregoing pages. 233

In most instances the feminine endings are given since those happen to be most commonly used in the genera already listed, but these are followed by the masculine and neuter terminals where they also have been used previously. The English is a free translation of what sometimes are obscure and occasionally even somewhat fanciful descriptions.

abrotanum	The old Latin name for Southernwood; *Artemisia*
absinthium	The old Latin name for Wormwood; *Artemisia*
absinthoides	Like *absinthium*
acanthifolia	With leaves like *Acanthus*; spiny
acanthium	Spiny or prickly
acaulis -e	Stemless or apparently stemless
acerosum	With stiff needles; or sharp
acetosella	Acid or sour
achilleifolia	With leaves like *Achillea*
acmopetala	With pointed petals
acmophyllus	With pointed leaves
aconitifolius	With leaves like *Aconite*
acris, acre	Biting or acrid
acuminata	Long pointed; usually the flower petals
acutifolia	With leaves shaped to a point
adscendens	Turning upwards; the flowers or leaves
adunca	Hooked or bent backwards

aesculifolia	With leaves like *Aesculus;* horse chestnut
aestivalis, aestivum	Of the summer; usually flowering
aethiopica	Of Ethiopia
affinis	Close to another species; allied
africanus	Of Africa
agavifolium	With leaves like *Agave*
aizoides	Like *aizoon,* but distinct
aizoon	Ever living or ever green
ajacis	Said to be founded on some marks at the base of the petals which are supposed to look like the letters AIAI (of *Delphinium*)
alata -um	With wings or flanges; usually the stems
alba -us -um	White
albicans	Whitish
albida	Turning or shading to white
albiflora -us	With white or whitish flowers
albo-coccinea	White and red
albo-maculata	With white spots; usually the leaves
albo-marginata	With a white edge; usually the leaves
albo-picta	As if splashed or painted with white
alcea	The old generic name for *Malva;* Mallow
alkekengi	A Japanese name
althaeoides	Like *Althaea;* Hollyhock

235

alpestris -tre	Found in the lower Alps
alpina -us -um	Found in the Alps
amabilis -e	Lovely; usually the flowers
amara	Bitter to the taste
ambigua	Of uncertain identity
ambrosiacum	With the fragrance of ambrosia
amelloides	Like *Aster amellus*
amellus	A name given by Virgil to a blue flower like an aster found growing by the River Mella in Northern Italy
amethystina -us -um	Of amethyst color
americana	Of America
amoena	Pleasing or lovely
amplexicaulis	Stem clasping; the leaf base clasping the stem
amurensis	Of Amur; Siberia
anchusifolium	With leaves like *Anchusa*
anemonoides	Like *Anemone*; the flowers
anethoides	Like *Anethum*; Dill
anglicum	Of England
angularis, angulosa	Having angles or corners, or with angular lobes
angustifolia -ium	With narrow leaves
annua -uus -uum	Annual
anomala	Unusual in relation to allied species
antirrhinoides	Like Antirrhinum
apennina -um	Of the Apennine Mountains; Italy

aperta	Opened or exposed; the flowers opening wide
aquilegifolium	With leaves like *Aquilegia*
arabica -us -um	Of Arabia
arachnoideum	Appearing to be covered with a spider web
arborea -eus, arborescens	Tree like; of tree-like growth
arctica	Of Arctic regions
arenarius -eum	Found growing in sand
argentea -eum	Silvery; with a sheen or lustre
argenteo-variegata	With silver markings; the leaves
argutifolius	With toothed leaves
aristata	Bearded or awned; like bearded corn
arizonica	Of Arizona
arkansanum	Of Arkansas
armeniacum, armenum	Of Armenia; South-west Asia
arvensis	Of the fields
asarina	An old generic name probably meaning gummy snouted
asclepiadea	Resembling *Asclepias;* Milkweed
asiatica -us	Of Asia
asperum	Rough; with minute points
asteroides	Like the *Aster*
asturiensis, asturicus	Of Asturia; now Northern Spain
atlanticum	Of the Atlantic regions

237

atrata	Darkened or blackened
atro	A prefix used before a color description indicating darker or darkened; as *atro-purpurea, atro-sanguinea, atro-violacea,* etc.
aubrietoides	Like Aubrieta
aurantiaca -us -um	Orange or orange yellow
auratum	With golden rays; the flower
aureo-marginata	With yellow margins; usually the leaves
aureo-variegata	With gold or yellow markings; usually the leaves
aurea -eus -eum	Golden
auricula	An ear; the leaves shaped like ears
australasica	Of the Australasian regions
australis	Of or from the south; or the Southern hemisphere
austriacum	Of Austria
autumnalis -ale	Of the autumn; usually flowering
azorica	Of the Azores
azurea -us -um	Azure blue
baicalensis	Of Baikal; Russia
balearica -um	Of the Balearic Islands; Majorca, Ibiza, etc.
balsamina	Balsam; an old generic name
basilicum	Basil; an old generic name
barbata -us	Bearded, bearing tufts of hair
belladonna	Pretty lady
bellidifolium	With leaves like *Bellis;* the Daisy

bellidiformis	In the form of *Bellis*
bellidioides	Like *Bellis*
bellum	Pretty
betonica	Betony
betonicifolia	With leaves like Betony
betonicoides	Like Betony
bicolor	Of two colors
bicornis	Having two horns; usually the seed pods
biennis	Bienniae; of two years
bifida	Split or divided into two
biflora -ius -ium	Having two flowers; or flowers in pairs
bifolia -ius -ium	Having two leaves; or leaves in pairs
bithynica	From Bithynia; now North-western Turkey
bipinnatus	The leaves doubly pinnate or feathered
blanda	Agreeable or pleasant
boliviensis	Of Bolivia
bombyciferum	Silky
borealis	Of the northern regions
botryoides	Like a bunch of grapes; the shape of the flower clusters
brachycalyx	With a short calyx
brachypus	With a short foot or base; usually the root stock
bracteata -us -um	Bearing bracts or modified leaves behind the flowers
brasiliensis	From Brazil
brevicaulis	With a short stem

brevipes	Short stalked
breviscapa	With a short scape or flower stem
bryoides	Like moss; of mossy growth
bulbifera	Having a bulbous rootstock; bearing bulblets
bulbocodium	With a fleecy covering to the bulb
bulbosa	Bulbous
bulgaricum	Of Bulgaria; the Balkans
bullata -us	With a puckered appearance; usually the leaf
byzantinum	Of Byzantium; Turkey
caerulea -eum	Sky or true blue
caerulescens	Tipped or tinted with blue
caesia -ius -ium	Bluish grey or grey; usually the foliage
caespititius	Growing in carpet-like patches
caespitosa -um	Growing in tufts
calabrica	Of Calabria; Italy
calathina	The flower shaped like a basket
calcarea	Chalk white; or found growing in chalky soil
calendulacea	Of the form or color of *Calendula*; pot marigold
californica -us -um	Of California
calliopsidea	Of the appearance of *Calliopsis* (see *Coreopsis*)
callosa -um	Having a hard skin
calycanthema	The calyx colored like the flower

cambrica	Of Wales
campaniflora	With bell-shaped flowers
campanularia	Describing bell flowers though
campanulata -us -um	species bearing this name may
	not belong to *Campanula*
campanuloides	Like *Campanula*
camtschatica	Of Kamchatka; Russian Far
	East
canadensis -ense	Of Canada
candicans	White; the flowers
candida -us -um	Pure white; the flowers
candidissima	Shining white; the flowers
canescens	Hoary or greyish white;
	usually the foliage
canina	With sharp teeth (thorns);
	sometimes also used in the
	sense of as common as
	dogs
cannabinum	Resembling *Cannabis*; Hemp
cantabrica	Of Cantabria; Northern Spain
capensis	From the Cape of Good Hope
capitata -us -um	Formed in a head; usually the
	flowers
cappadocica	Of Cappadocia; Central Turkey
capreolata	Having tendrils
cardinalis -e	Colored like a cardinal's robes;
	scarlet
carinatum	Keeled like a boat; the form of
	the seeds
carinthiaca	From Carinthia; Austria
carmineus	Carmine
carnea -um	Flesh colored; the flowers

241

carolina, *caroliniana*	Of Carolina
carpatica	From the Carpathian Mountains; Eastern Europe
carthusianorum	Of the Carthusian monks or monasteries
caryophyllus	Obscure; literally not leaved; but also the old name of a shrub which yields the spice clove transferred to *Dianthus* and probably referring to the clove-like perfume of these flowers
cashmeriana	Of Kashmir; Northern India
cathayanum	Of Cathay; the old name for China
caucasica -icus -icum	Of the Caucasus; Asia
caudatus	Bearing a tail; usually the shape of the inflorescence
caulescens	With long stems
cautleoides	Like *Cautleya;* a plant of the *Zingiber* (Ginger) family
cerastioides	Like *Cerastium;* Snow in summer
chalcedonica	Of Chalcedonia; Asia Minor
chamaedrys	An old name for Germander meaning on the ground
chamaejasme	A dwarf or low on the ground jasmine
cheilanthifolia	With leaves like *Cheilanthus;* Lace fern

cheiri	Probably Latinised from the Arabic *kheri*, a fragrant red flower
chilensis *chiloense*	1. Of Chile, 2. of the Island of Chiloe, Chile; but either form is used
chinensis	Of China
chrysantha -us -um	Golden flowered
chrysanthemoides	Like *Chrysanthemum*
chrysocoma	With yellow tufts or hairs
chrysographes	With yellow or golden veins or markings
ciliata *ciliosum*	Fringed like an eyelash
cilicia -ium	Of Cilicia; Southern Turkey
cinerea -eum	Ashen grey; usually the foliage
citrina -um	Lemon yellow
citriodorus	Lemon scented
clematidea	Like a small clematis
clematiflora	With flowers like the clematis
clivorum	Of the hills
Cneorum	An ancient Greek name; the meaning obscure
coccinea -eus -eum	Bright deep pink
cochlearifolia	With spoon-shaped leaves
coeli-rosa	Rose of Heaven
colchiciflora	Flowers like *Colchicum*; Autumn crocus
columbaria	An old name meaning dove-colored
communis	Growing in communities

243

commutatum	Changing in form or color
comosa -us -um	With tufts of hair
compacta -um	Growing in a compact form
conferta -um	Crowded together; the habit of
congesta	growth, sometimes the flowers
consolida	Joined in one; a flower spike
convolvulacea	Twining; like *Convolvulus*
cordata -um	Heart shaped; usually the leaves
cordifolia -us	With heart shaped leaves
coreanum	Of Korea
coridifolium	With leathery leaves
cornuta	Bearing horns or spurs; usually the flowers
coronata -aria -arium	Crowned or wreathed
corsica -us -um	Of Corsica, the Mediterranean
crassifolia	With thick leaves
cretica -um	Of Crete, the Mediterranean
criniflorum	With hair-like flowers or petals
crinita -us	Long haired or fringed
crispa -us	Closely curled or crested
cristata	Crested or comb-like
crocata	Saffron yellow
cruenta	Blood red
cucullata -aria	Hooded, the flowers; the upper petals bent over
cupreus	Of a coppery color
cuspidata	Tipped with a firm point; usually the leaves
cyanea -us	Dark blue

cyclamineus	Of the form of *Cyclamen*; the flower
dalmatica -us	Of Dalmatia; now Yugoslavia
damascena	1. Of Damascus, 2. Damask (the same origin)
dasycarpa	With woolly seed heads or fruits
dealbata	Whitened as with powder
debilis	Weak or small relative to other species
decapetalus	Having ten petals
decorata	Decorated or handsome
decurrens	Running down; the leaf blade extends as a wing or ridge down the stem
decussata	Cross shaped; as when the leaves are borne in alternate opposite pairs
deltoidea -es	Like the Greek letter delta; triangular flower petals
demissa	Hanging down
dens-canis	Dog's tooth
densiflorus	With dense or closely set flowers
dentata	Toothed like a saw
denticulata	With fine teeth, usually the leaf edges
depressa	Appearing to be pressed down flat
dictamnus	An old generic name for Dittany

didyma	Double or twin
diffusum	Spreading loosely
digitata	Lobed like fingers; the leaves
dilatum	Widened or expanded; the leaves
dinarica -us	Of the Dinaric Alps; Yugoslavia
diphylla -us -um	With two leaves or leaflets; arranged in pairs
dipterocarpum	The seed pods or fruits having two parts
dissitiflora	The flowers far apart; loosely arranged
disticha	Arranged in two rows; usually the flowers
divaricata *divergens*	Spreading apart; the habit of growth
diversifolia -us	With different shaped leaves
doronicum	An old generic name
dracunculus	Like a dragon
dubia -ium	Of doubtful origin
echioides	Like *Echium;* Viper's bugloss
edulis	Edible
elata -um *elatior*	Tall or taller than other species
elegans	Elegant or graceful
emodi	Of Mount Emodus, Himalayas
enneaphylla	With nine leaves or leaflets; nine divisions to the leaf
ensifolia	Sword shaped, the leaves
ephemerum	Short lived
erecta	Upright

246

erinus	An old generic name meaning early
esculenta	Edible
europeus -eum	Of Europe
exaltata	Erect, tall or commanding
eximia	Exceptional or distinguished
farinacea	Dusted with farina or meal
fasciculata	In tight clusters or bundles
fastigiata	With erect clusters of twigs or stems
fastuosum	Stately or bountiful
ferruginea	Rusty red; usually the flowers
ficaria	Like *Licus;* the Fig
ficifolia -ium	With leaves like *Ficus;* Fig
filamentosa	With threadlike filaments
filifolia -ium	Fine or threadlike foliage
filiformis	Of fine or threadlike form or growth
filipendula *filipendulina*	Joined or hanging by threads; usually the roots
fistulosa	Hollow or tubular; usually the stalks
flava -um	Pure yellow
flore-pleno	With double flowers
floribunda	Flowering profusely
flos-jovis	Flower of Jove; a name of ancient origin
foetida -us	Having an unpleasant odour
foliosa	Having abundant foliage
formosa -um *formosana*	Handsome or beautiful

formosissima	Superlatively beautiful
fragiformis	Of the form of *Fragaria*; Strawberry
fraxinella	Resembling *Fraxinus*; Ash; usually the leaves
frigida -um	Stiff
frondosa	With much leafage
frutescens *fruticosa*	Shrubby
fulgens *fulgida*	Shining or glowing; usually scarlet
fulva	Of a tawny color
fumariifolia	With leaves like *Fumaria*; Fumitory
gallica	Of France
galacifolia	With leaves like *Galax*; Wand plant
garganica	Of Monte Gargano, Italy
genevensis	Of Geneva
gentianella	The diminutive of *Gentian*; a little gentian
gentianoides	Like *Gentian*, usually the flower color
gigantea -eum	Large in relation to other species
glabra, glaber, *glabrata, glaberrima*	Smooth or hairless
glacialis	From icy or high alpine regions
glauca -um *glaucescens*	Greyish or bluish green
glauciifolia	With greyish or bluish foliage

globosa	Globe shaped; usually the flower head
glumaceum	Having chaffy bracts
glutinosum	Sticky
gracilis	Graceful or slender
gracilipes	
graminifolia -us	With foliage like grass
grandicephalum	With a large head
grandiflora -us -um	With larger flowers than other species
grandis	Large or impressive
granulata	Having a granulated appearance; the small root tubers
haemantha	With blood red flowers
haematodes	Blood colored
hederacea	Growing like *Hedera;* Ivy
hederaefolium	With leaves like *Hedera;* Ivy
helenium	An old generic name; after Helen of Troy
helianthoides	Like *Helianthus;* Sunflower
helvetica	Of Helvetia; Switzerland
heracleifolia	With leaves like *Heracleum;* Cow parsnip
heterophylla -us	With leaves of differing or unusual shape
hexapetala	Having six petals
himalaicus	Of the Himalayan regions
hippomanica	From the place of the horses; i.e. the meadows
hirsutus	Roughly hairy
hirta	With smaller hairs

hispanica -us -um	Of Hispania; Spain
hispida	Rough, with bristly hairs
histrioides	Like an actor; the flower being gaily adorned
horminum or *horminium*	An old generic name of doubtful origin and meaning
horridula	Offensive or unpleasant; sometimes bristly
hortensis	Of horticulture; originating in gardens
humeana	Growing on the surface of the ground
humilis	Of humble growth
hybrida -us -um	Hybrid
hyemalis	Of the winter
hypochondriacus	Of a sombre appearance
ibericum	Of Iberia; the ancient name for Spain
iberideum	Resembling *Iberis*; Candytuft
iberidifolia	With leaves like *Iberis*
ida-maia	Obscure, but probably a woman's name Latinised; Ida May
illyricus	Of Illyria; an old name, now Yugoslavia
imperialis	Majestic or powerful
incana -us -um *incanescens*	Greyish or hoary; usually the leaves
incarnata	Flesh colored or flushed with pink
incisifolia	With deeply cut leaves
incomparabilis	Incomparable

indica -um	Of India
inodora	Without a scent
insignis	Significant or distinguished
integerrima	With a smooth edge
integra	Whole, entire; of uncut or
integer	undivided leaves
integrifolia	With a smooth edge; the leaves
involucrata	With edges rolled inwards
iridiflora	With flowers like *Iris;* or of a shining color
italica -um	Of Italy
japonica -um	Of Japan
jalapa	An old generic name; Jalap
kewensis	Of Kew; the Royal Botanic Gardens, London
labradorica	Of Labrador
laciniata -us -um	Fringed or slit; usually the petals but sometimes the leaves
lactiflora	With milky white flowers
laevigata -us	Smooth or slippery
laevis	
laetiflorus	With flowers of a gay or joyful appearance
lanata -um	Covered with long woolly hair
lanuginosa	
lancastriense	Of Lancaster, England
lanceolata	Shaped like small lances; the leaves
lasiocarpa -us -um	Having woolly seed heads or fruits

251

lathyris	An old generic name for Caper spurge
latifolia -us -um	Having wide leaves
lavandulaceus	After the form of *Lavandula*; Lavender
lavandulifolia	With foliage like *Lavandula*; Lavender
laxa	Growing loosely
laxiflora	With flowers in loose clusters
leptosepala	With fine or narrow sepals
leucantha	With white flowers
leucanthemum	
leucocarpa	With white berries
libanotica	Of Lebanon; Eastern Mediterranean
ligtu	A horticultural name
lilacina	Of the color of Lilac
liliago	A silvery lily
liliastrum	Star lily
linearifolia	With narrow parallel sided
linearis	leaves
lingua	A tongue; usually the shape of the leaves
linifolia -ius -um	With leaves like *Linum*; Flax
lobata -us	Divided into lobes; usually the leaves
longiflora	With long flowers
longifolia -ius	With long leaves
longiscapa	With long scapes; the flower stalks
longissima	Longer than usual, usually the flower spurs

lucida	Clear or shining, the color of the flowers
lupinoides	Like *Lupinus*; Lupin
lutea -eus	Yellow
luteola *lutescens*	Yellowish to green or buff
lycoctonum	Wolf's bane
macedonica	From Macedonia; northern region of the Balkan Peninsular
macrantha -us	With large or long flowers
macrocarpa -us	With large fruits or seed pods
macrocephala	With a large head; usually the flower
macropetala	With large petals
macrophylla	With large leaves
macrosiphon	With a long tube or spur; usually the flowers
maculata -um	Spotted or blotched; the flowers or leaves
magellanica -us	Of the Straits of Magellan; Chile
magnifica -um	Imposing
majalis	Flowering in May
major, *majus*	Greater or larger
malacoides *malvaeflora*	Flowers like *Malva*; Mallow
manicata	Sleeved; usually applied to a loose covering or sheath
marginata	Margined with another color; the leaves

253

maritima -um	From sea coast areas
maroccana	Of Morocco; North Africa
matronalis	Of the dame or matron
mauretanica -icus	From Mauretania; now Morocco
maxima -um	The greatest or largest
medius -ium	1. Intermediate between two types; 2. Middle sized
meleagris	The old Greek name for a guinea fowl; speckled
melissophyllum	With leaves like *Melissa*; Common balm
menthaefolia	With leaves like *Mentha*; Mint; usually the scent
metallica	Having a metallic sheen; usually the leaves
mexicana -um	Of Mexico
Mezereum	From the old Persian *mazaryum*; thought to signify poisonous or deadly
micrantha	With very small flowers
microcarpa	With small fruits or seed pods
microlepis	With small scales; or small and neat
micropetala	With small petals
microphylla	With small leaves
millefolium	With many leaves or leaf segments
minor	Smaller than other species
minima -us minus	The smallest
minuta	Very small; minute

missouriensis	Of Missouri or the Missouri River
moldavica	Of Moldavia; Northern Black Sea area
mollis	Smooth; or with soft velvety hair
moly	An old Greek name for one species of Allium
monstrosum	Of abnormal appearance or development
montana -um	Of the mountains
moschata -us	Having a musky perfume
mucronatus	With small points; usually the leaves
multicaule	With many stems
multicolor	Of many colors
multiflora -um	With many flowers
mutabilis	Changing in form or color
muralis	Growing on walls
myosotidiflora	With flowers like *Myosotis;* Forget me not
nan -us	Dwarf
nana-compacta	Dwarf and compact; neat
nankinensis	Of Nankin; China
napellus	Turnip rooted
narcissiflora -um	With flowers like Narcissus
narbonensis -ense	Of the Narbonne; France
natalensis	Of Natal; South Africa
neapolitanum	Of the area about Naples; Italy
neglectus	Neglected or overlooked; usually an insignificant plant without its flowers

nemorosa, nemoralis *nemophila*	Growing in woodland glades
nepalensis *nepaulensis*	Of Nepal; Northern India
nervosa -um	Having distinct veins or nerves; usually the leaves
niger, nigra *nigrum*	Black
nipponica	Of Japan
nitida	Lustrous, shining or smooth
nivalis	From near the snow line
nivea	Snow white
nobile -is	Of stately or noble appearance
nonscriptus	Not written of or described
nootkatensis	Of Nootka Sound; North America
novae-angliae	Of New England
novi-belgii	Of New York; a name of historical origin
novae-zealandiae	Of New Zealand
nudicaule -is	Having naked or leafless stems
nummularia -ium	Like a coin; round; usually the shape of the leaf
nutans	Nodding or drooping; the flowers
nyctaginiflora	Obscure; but probably flowering or fragrant by night
obcordata -um	Inversely heart shaped; the leaves
obliqua	Slanting sideways or having unequal sides

obtusa	With a blunt apex
ochroleuca -us -um	Yellowish white or pale yellow
octopetala	Having eight petals to the flower
oculis-Christi	The Eye of Christ; of the flower
ocymoides	Like *Ocymum;* Sweet basil
odora -us	Fragrant; sweet smelling
odorata -us -um	
officinalis -e	Of the shop, the herbalist's;
officinarum	of value and service to man
oleracea	Aromatic
olympica -us -um	Of Olympia; Greece
oppositiflora -us	With flowers opposite to each other
oppositifolia -ius -ium	With leaves opposite to each other on the stem
orbiculatum	Disk or ball shaped, or coiled into a disk
oregana	Of Oregon
oreodoxa	A beauty of the mountain
orientalis -e	From the East
ornata	Ornamental, handsome
ovata	Egg shaped; usually the leaves
oxypetala	With pointed petals
oxysepala	With pointed sepals
pacifica	From the Pacific Ocean regions
pallidiflora	With pale colored flowers
palmata -um	The leaves lobed like a hand
palustris	Of the marshes; growing in wet places

paniculata	With the flowers carried in panicles
papilio	Resembling a butterfly; the flowers
pardalinum	Spotted like a panther or leopard
pardanthina	With spotted or blotched flowers
parnassi	Of Mount Parnassus, Greece
parnassifolius	With leaves like *Parnassia*; Grass of Parnassus
parviflora	With small flowers
patens	Spreading
patula	Somewhat spreading; or spreading open, the flowers
pauciflora	With few flowers
paucifolia	With little foliage
pavonia	Of the peacock; as showy as a peacock
pectinata	Obscure; but literally in the form of a comb
pedata	Shaped like a bird's foot (fanciful)
pedunculata	With a distinct stalk
pelargoniflorum	Having flowers like *Pelargonium*
peltata -um	Shield shaped; usually the leaves but sometimes referring to the habit and appearance of some high alpine plants
pendula -um	Hanging down, swinging; usually the flowers
peregrina -um	Wandering; or straggling in growth

perennis -enne	Perennial
perfoliata -um	The leaf blade surrounding the stem; the stem appearing to pass through the leaves
persica -um	Of Persia
persicifolia	With leaves like *Persica;* Peach
peruvianum	Of Peru
philadelphicus	Of Philadelphia
phoenicea -eum	Colored like the Phoenix; red or scarlet
physaloides	Like *Physalis;* Chinese lantern plant
picta	Colored as if painted
pilosa	Lightly covered with soft hair
pinnata -us -um	Feathered; pinnate or compound leaves with leaflets in parallel pairs
plantaginea -eum	Like *Plantago,* Plantain
platypetala	With broad flat petals
plena -um	Double
plumbaginoides	Like *Plumbago*
plumosa *plumarius*	Plumed or feathery
pluvialis	Of the rain; flowering in the rainy season
podophylla	With stalked leaves
polychroma	Of many colors
polypetala	With many separate petals
polyphylla -us -um	With many leaves or leaflets
pontica	From the shores of the Black Sea
praecox	Developing early, or earlier than other species

praestans	Outstanding or excellent
pratensis -ense	Growing in meadows
primula	Primrose colored
primulina	
primuloides	Like the *Primula*
procera	Tall or slender
procumbens	With trailing, prostrate stems
prolifera -us	Prolific; growing or bearing profusely
prostrata -um	Lying flat, prostrate
pseudo	False; False armeria, False narcissus, etc.
psittacina -us	Colored like a parrot
ptarmica	Causes sneezing
pterocephala	A winged head; the form or appearance of the flower
pubescens	With soft downy hair
pulchella -us -um	Beautiful
pulcher	
pulcherrima -um	Most beautiful
pulchra	Pretty
pulverulenta -um	Powdered, as if with dust or meal
pumila -us -um	Small in relation to other species
pumilio	
purpurascens	Turning purple, purplish, or soft purple
purpurea -eum	True purple
purpureo-auratus	Purple and gold
purpureo-caeruleum	Purple and blue
pusilla	Small; weak or insignificant
pycnostachya	With a dense spike; the flower head

pygmaea	Of dwarf stature
pyrenaica -um	From the Pyrenees; northern Spain
quamash	A North American Indian name for *Camassia*
racemosa	With flowers carried in racemes
radiata	Spreading out like rays; usually the petals or florets
ramosum	Much branched
ranunculoides	Like *Ranunculus*; Buttercup
recta	Growing upright
recurva	Bent over and downwards
recurvifolia	With the leaves curving backwards
rediviva	Reviving; the plant appearing to be dead but suddenly flowering
reflexa	Turned back; usually the flower petals
regale, regia	Regal
reniformis	Kidney shaped; usually the leaves
repandum	Scalloped or waved; usually the leaf margins
repens	The stems creeping and rooting
reptans	
reticulata	Netted; usually the leaf venation but sometimes the outer covering of bulbs
retusa -us	Rounded and/or notched; the petals or leaves

revolutum	Rolled back from the margin or apex; the petals or leaves
rex	Royal
rhoeas	An old generic name; of *Papaver rhoeas*, Corn poppy
rhodopensis	Of the Rhodope Mountains; Greece
rigens	Rigid
rigida	
ringens	Opening wide; the flowers
ritro	A Southern European local name
rivale	Growing by brooks and rivers
rivularis	
robusta -us	Strong, robust
rosea -eus -eum	Rose colored; pink
rosaeflora	With flowers like the rose
rotundifolia -ium	With round leaves
rubra	Red
ruber	
rugosa -um	With wrinkled leaves
rupestris -re	Found growing among rocks
rupicola	Found growing on ledges or cliffs
saccharata	Sweet
salicaria	Like *Salix*; Willow; but referring to *Epilobium*, Willow herb
salicifolius -ium	With leaves like *Salix*; Willow
salsoloides	Like *Salsola*; Saltwort
sambucifolia	With leaves like *Sambucus*; Elder

sancti-johannis	St. John
sanguinea -eus -eum	Blood red
sardensis	Of Sardinia, Mediterranean; or Sardis, Turkey
sarmentosa	With long slender runners
sativus	Cultivated as a crop plant
saxatilis -e	Growing among rocks
saxosa	
saxicola	Of the rocks
scabra -er -um	Rough to the touch
scandens	Climbing
schafta	Of obscure origin; but probably a place name in the Caucasus
scilloides	Like *Scilla*; Squill
sclarea	Clear; possibly alluding to the use of plants so named in eye lotions
scoparia	Brush or broom like
scordium	An old name for *Teucrium*; Germander
segetum	Of cornfields
semperflorens	Ever flowering
sempervirens	Ever green
septemfida	With seven clefts; usually the flowers
serbica	Of Serbia; the Balkans
serpyllifolia -ius -ium	With leaves like *Thymus serpyllum*; Wild thyme
serpyllum	An old Greek name for Wild thyme
sessile	Stemless; the flowers

263

sibirica -um	Of Siberia
signata	1. Notable
	2. Marked as with writing in another color
sikkimensis	Of Sikkim; Southern Himalayas
silvatica *silvestris*	Of or from the woods
simplicifolia	With simple (entire or undivided) leaves
sinensis -e	Of China
sino-ornata	Chinese and ornate
sinuata -um	Having sinuous or wavy margins; usually the leaves but sometimes the flowers
slavica	Of Slavia; an area of Central Europe
somniferum	Bringing sleep
spathulata *spathularis*	Spoon shaped, usually the leaves
speciosa -us -um	Showy
speciosissimum	Exceptionally showy
spectabilis -e	A spectacle; striking
speculum	A mirror
sphaerocephalus -um	A round head; the flower
spica	A spike; the flower head
spicata -um	Arranged in spikes; the flowers
spinosa -us	Bearing spines
splendens	Bright, shining, splendid

splendida -um

spurium	False or doubtful; a species so called sometimes has other names
squamigera	Bearing scales
stellata	Star shaped
stolonifera	With stolons or runners
striatum	Channelled or grooved; usually the leaves
strumosa	Having tubercles
suaveolens	Sweet smelling
subulata	Pointed and sharp like an awl; the leaves
succisa	Appearing bitten or broken off
suffruticosa	Growing part herbaceous and part shrubby
sulphurea -eus	Sulphur colored
sultanii	After a Sultan of Zanzibar
superba -um	Superb
superbissima	A superlative of superb; possibly a horticultural invention
sutchuenense	Of Sutchuen Province; China
sylvatica *sylvestris*	See silvatica, silvestris
syphilitica	Alluding to the disease; for which plants so named were once thought to be a cure
tabularis	From Table Mountain, Cape of Good Hope
tagetifolia	With leaves like *Tagetes*
tanacetifolia	With leaves like *Tanacetum*; Tansy

265

tangutica	Of Tangshan; China
tarda	Late developing or flowering
tardiflora	
tasmanica	Of Tasmania; Australia
tatarica	Of Tartary; now South-east Europe and Asia
tauricum	Of the Taurus Mountains; Turkey
tectorum	Found growing on roofs
tenella	Delicate, light or soft
tenuicaulis -e	With fine stems
tenuiflora	With fine or delicate flowers
tenuifolia -ius -ium	With fine leaves
tenuior	Finer or more delicate than other species
tenuissimus	Extremely fine or slender
tetra	1. Offensive; 2. Of four parts
texensis	Of Texas
thalictrifolia	With leaves like *Thalictrum*
thapsus	Of Thapsus in ancient Africa (now Tunisia); or of Thapsos, Greece
thyrsoides	The flower head like a thyrse
tinctoria	A plant used for dyeing
tingitanus	Of Tangier; North Africa
tomentosa -um	With a covering of short hair
transsilvanica	Of Transylvania, Central Europe
triandrus	Having three stamens
trichocalyx	With a hairy calyx
trichophylla	Fine or hair leaved
trichosantha	Hairy flowered

tricolor	Of three colors
trifida	Divided into threes
triflora	With flowers in threes
trifoliata	With leaves or leaflets in threes; or leaves with three lobes
triloba	With three lobes; usually the leaves
trimestris	Maturing or flowering in three months
trionum	Of three colors or zones
triquetra -us -um	Three angled or cornered; the stems
tristis	Sad; generally the flower color
tuberosa -us -um	Bearing or resembling tubers; the roots
tuolumnense	Of the Tuolumne River area, California
tupa	An old name of Chilean origin for one species of *Lobelia*
turkestanica	Of Turkestan
uliginosum	Moisture loving; growing in marshes
umbellata -us -um	Bearing the flowers in umbels
umbrosa -um	Found in shady places
unalashkensis	Of Unalaska; one of the Aleutian Islands
uncinatum	Hooked; with barbed bristles
undulatum	Undulating or waved at the margin; usually the leaves
unguicularis	Narrow clawed; the base of the petals

267

uniflora -us -um	Bearing one flower; each stem
urticaefolium	Nettle leaved
usitatissimum	Most commonly used by man
uvaria	An old generic name for *Kniphofia* meaning clustered
uva-ursi	Grapes and bear; the bear's grapes
vaccaria	An old generic name meaning "cow herb"
vacciniifolium	With leaves like *Vaccinium*; Cranberry
variabilis	Not constant in appearance; usually the color
variegata -us -um	Marked with another color; usually the leaves
venosa	Conspicuously veined
venusta -us -um	Pleasing or lovely
vera -is	True; usually when a species is recognised as being distinct rather than a form of another
vernalis -e *verna -us -um*	Of the spring
versicolor	Variable, or of more than one color
villosa -us -um	With long loose hairs
viminea	Bearing long flexible stems
vinciflora	Flowers like *Vinca*; Periwinkle
violacea	Violet color
virgatum	Twiggy in growth
virgaurea	More correctly virga-aurea; an old name meaning golden twig
virginica	Of Virginia

viridiflora	With green flowers
viridipicis	With green tips or points
viridis -e	Green
viscaria	Sticky to the touch
viscosa	
viscida	Exuding a sticky secretion
viticella	A bower of vines
vitifolia	With leaves like *Vitis;* Vine
volubile	Twining
vulgaris -e	Common, ordinary
xanthina	Golden yellow
xanthospila	With yellow spots or patches
yunnanensis	Of Yunnan; China
zalil	A native Afghanistan name
zinniiflora	With flowers like *Zinnia*

A Short List of Typical Personal and Commemorative Names

These usually may be recognised by their terminals. Masculine forms end in *-ii* if the last letter of the word stem is a consonant, or *-i* if the stem ends with a vowel, -y, or -r. The feminine endings are *-ae*, or *-iae*, and strictly should only be applied as indicating a connection with the discovery of the plant; although in fact they appear often to be used as honor or complimentary names. Those commemorating a dead person or one clearly not associated with the actual finding of the plant end in *-ian* plus the gender terminal of the genus, as *-ianus* when this is masculine, *-iana* when it is feminine, and *-ianum* for neuter; but the first letter, *i*, may be dropped when the word stem ends with e.

alexandrae	Named for a Princess Alexandra.
bigelovii	Professor Bigelow of Boston; U.S.A.
clusiana	After Carolus Clusius who first cultivated the tulip at Leyden, Holland, in the late sixteenth century.
davidii	Père Jean Pierre David, a nineteenth century Jesuit priest who is said to have sent back

to France more than 2,000 species of plants from China.

douglasii David Douglas, a nineteenth century Scots plant collector who was sent by the then Horticultural Society (later the Royal Horticultural Society) to collect trees and plants in North America.

farreri Reginald Farrer, author, plant collector and authority on alpine plants who—among many other journeys—explored the Kansu-Tibetan border country in 1914 with William Purdom of the Arnold Aboretum, Massachusetts.

haageana After Haage, a famous German nurseryman.

lamarckiana After Jean Baptiste Lamarck, the great French naturalist and zoologist of the eighteenth and nineteenth centuries.

luciliae A Mme. Lucille Boissier.

olgae A Princess Olga of Rumania.

sternbergii Count Caspar Moritz von Sternberg, a botanist of Prague in the eighteenth and nineteenth centuries.

tubergenianum After the famous family of Dutch bulbsmen, van Tubergen.

271

wilsoniana	After E. H. Wilson (1876–1930), a plant collector of Birmingham, England, who later became a director of the Arnold Aboretum. He travelled largely in China and, among 1,000 or so other plants, discovered *Lilium regale*.

The Family Groups

Genera which appear to be related by like characteristics are grouped into larger families, and these names are recognisable as ending in *-ae* or *-eae*. These may describe a simple and obvious similarity between the genera, like *Compositae, Cruciferae, Umbelliferae,* etc., or they may be derived from a type or main genus with the family—as *Scrophularia* (Figwort) which is of relatively little garden importance but gives its name to *Scrophulariaceae,* the family containing *Antirrhinum.*

How many families of flowering plants there are and exactly how the genera are to be grouped is still a matter of some controversy among different schools of botany; the number appears to vary from 200 to 354—or according to one authority 411—and as methods of classification grow more precise a genus or genera which have been regarded as belonging to one family may be removed to others or even given distinct groups of their own. The following list therefore includes only a few of the more important families as being of most interest to the gardener in showing the relationships of apparently very dissimilar plants; as *Buttercup, Clematis* and *Delphinium* all belong to the family *Ranunculaceae,* and *Onion, Colchicum* and *Erythronium* are *Liliaceae*—although, as an instance of the uncertainty of classifications, it may be noted that some botanists place *Onion* and several other members of the *Lilium* family with *Amaryllis.*

The genera shown here are only those which appear in the main part of this book and each family will of course contain many more than are included.

AMARYLLIDACEAE: from the type genus *Amaryllis*

Alstroemeria	Ixiolirion	Nerine
Amaryllis	Leucojum	Polianthes
Galanthus	Lycoris	Sternbergia
Hymenocallis	Narcissus	Zephyranthes
		etc.

BORAGINACEAE: from the type genus *Borago* (Borage)

Anchusa	Echium	Myosotis
Arnebia	Heliotropium	Omphalodes
Borago	Lithospermum	Pulmonaria
Brunnera	Mertensia	Symphitum
Cynoglossum	Myosotidium	etc.

CAMPANULACEAE: from the type genus *Campanula,* bell flowers

Campanula	Lobelia	Specularia
Codonopsis	Michauxia	Symphyandra
Cyananthus	Phyteuma	Wahlenbergia
Edraianthus	Platycodon	etc.

Lobelia may sometimes be seen placed in Lobelia-

ceae

COMPOSITAE: plants with composite flowers (as Chrysanthemum, Dahlia, etc.)

Achillea	Dimorphotheca	Liatris
Ageratum	Doronicum	Ligularia
Amellus	Echinacea	Matricaria
Ammobium	Echinops	Onopordon
Anthemis	Erigeron	Pyrethrum
Arctotis	Eupatorium	Raoulia
Artemisia	Felicia	Rudbeckia
Aster	Gaillardia	Sanvitalia
Bellis	Gazania	Senecio
Brachycome	Grindelia	Silphium
Buphthalmum	Helenium	Solidago
Calendula	Helianthus	Stokesia
Callistephus	Helichrysum	Tagetes
Catananche	Heliopsis	Tithonia
Centaurea	Helipterum	Ursinia
Chrysanthemum	Hieraceum	Venidio-arctotis
Cineraria	Inula	Venidium
Cladanthus	Lasthenia	Xanthisma
Coreopsis	Layia	Xeranthemum
Cosmos	Leontopodium	Zinnia
Dahlia	Leptosyne	etc.

CRUCIFERAE: plants having flowers with four equal petals arranged crosswise

Aethionema	Erysimum	Lunaria
Alyssum	Hesperis	Malcomia
Arabis	Hutchinsia	Matthiola
Aubrieta	Iberis	Parrya
Cheiranthus	Ionopsidium	Thlaspi
Dentaria	Lobularia	etc.

IRIDACEAE: from the type genus *Iris*

Crocosmia	Iris	Sisyrinchium
Crocus	Ixia	Sparaxis
Dierama	Lapeyrousia	Streptanthera
Gladiolus	Montbretia	Tigridia
Hermodactylus	Schizostylis	etc.

LABIATAE: plants of which the corolla or calyx is divided into two parts suggesting lips

Ajuga	Monarda	Salvia
Dracocephalum	Monardella	Scutellaria
Hyssopus	Nepeta	Stachys
Lavandula	Oreganum	Teucrium
Melittis	Perilla	Thymus
Mentha	Physostegia	etc.
Moluccella	Rosemarinus	

LILIACEAE: from the type genus *Lilium*

Agapanthus	Fritillaria	Nomocharis
Allium	Galtonia	Ornithogalum
Anthericum	Helonias	Paradisea
Brodiaea	Hemerocallis	Polygonatum
Calochortus	Hesperoyucca	Puschkinia
Camassia	Hosta	Reineckia
Cardiocrinum	Hyacinthus	Scilla
Chionodoxa	Ipheion	Smilacina
Colchicum	Kniphofia	Tricyrtis
Convallaria	Lilium	Trillium
Endymion	Maianthemum	Tulipa
Eremurus	Medeola	Veratrum
Erythronium	Muscari	Yucca
		etc.

Agapanthus, ⎫ May sometimes be seen placed
Allium, Brodiaea ⎭ in AMARYLLIDACEAE

Hesperoyucca ⎫
Yucca ⎭ Sometimes in AGAVACEAE

MALVACEAE: from the type genus *Malva,* Mallow

Althaea	Kitaibelia	Malva
Callirrhoë	Lavatera	Sidalcea
Hibiscus	Malope	etc.

PAPAVERACEAE: from the type genus *Papaver,* Poppy

Argemone	Meconopsis	Sanguinaria
Eschscholtzia	Papaver	Stylophorum
Hunnemannia	Platystemon	etc.
Macleaya	Romneya	

PRIMULACEAE: from the type genus *Primula*

Anagalis	Cyclamen	Polyanthus
Androsace	Dodecatheon	Primula
Auricula	Lysimachia	Soldanella
		etc.

RANUNCULACEAE: from the type genus *Ranunculus,* Buttercup

Aconitum	Clematis	Pulsatilla
Adonis	Delphinium	Ranunculus
Anemone	Eranthis	Thalictrum
Aquilegia	Helleborus	Trollius
Caltha	Hepatica	etc.
Cimicifuga	Nigella	

277

ROSACEAE: from the type genus *Rosa;* Rose

Acaena	Filipendula	Rosa
Alchemilla	Geum	Sanguisorba
Dryas	Potentilla	etc.

SAXIFRAGACEAE: from the type genus *Saxifraga;* Saxifrage

Astilbe	Kirengeshoma	Saxifraga
Bergenia	Parnassia	Tiarella
Heuchera	Peltiphyllum	etc.
Heucherella	Rodgersia	

Parnassia   May sometimes be seen placed in PARNAS-SIACEAE

SCROPHULARIACEAE: from the type genus *Scrophularia,* Figwort

Alonsoa	Erinus	Penstemon
Antirrhinum	Linaria	Synthyris
Calceolaria	Mazus	Verbascum
Chelone	Mimulus	Veronica
Collinsia	Nemesia	Wulfenia
Digitalis	Ourisia	etc.

A Short Glossary of Botanical Terms

ANTHER. The male part of a flower which bears the pollen grains. It is usually attached to the flower by a stem known as the filament, and this together with the anther is called a stamen.

AWN. A thread or bristle attached to a part of the plant; usually to the fruit or seed.

BRACT. A modified leaf on the flower stalk or forming part of the flower head. It may be leafy and green, or colorful as with *Euphorbia pulcherrima,* Poinsettia. Bracts are often mistakenly referred to as flowers.

CALYX. The collective name for the sepals of a flower; the bract-like growths which protect the bud before it opens.

CAPSULE. A dry fruit formed from two or more carpels which split to shed the seed.

CARPEL. The collective name for stigma, style and ovary, the seed-bearing organ of flowers.

COMPOSITE. A flower which is actually a particular kind of inflorescence made up of many separate flowers or florets, each one complete in itself though modified in appearance and all united in a head. The characteristic of the family *Compositae*.

COROLLA. That part of the flower made up of true petals—usually the most conspicuous—surrounded by the circle or collar of sepals (calyx) which enclosed the bud.

CORONA. A term usually describing the cup or trumpet of the Narcissus, but often applied to any floral structure which may separate the corolla of a flower from its anthers.

CORYMB. A flat topped type of flower head composed of a series of florets borne on individual stalks and forming a rounded and more or less level head of blossom.

CRUCIFER. Bearing a cross. Referring to flowers of four petals arranged in a cross formation. The characteristic of the family *Cruciferae.*

DIGITATE. A term used to describe leaves which are arranged in the form of a hand, being united to a common stalk, such as those of Lupin.

FILAMENT. The lower part of the stamen which is usually slender or stalk like and bears the anther or pollen producing part of the plant. Where the filament is missing in some flowers the anther is described as sessile.

FLORET. One of the small individual flowers which makes up the head or inflorescence in the family *Compositae*. The central cushion of the head consists of disk florets and the outer frill of ray florets.

HEAD.   A short, dense cluster of flowers; or flowers that grow together in a globular fashion.

INVOLUCRE.   A whorl of bracts found immediately behind the flower in plants of the family *Compositae.*

KEEL.   The front oval shaped part of the flower in leguminous plants as distinct from standards and wings, the back and side petals respectively.

LEGUME.   The term to describe a seed pod such as those of peas and beans; a member of the pea family *Leguminosae.*

MONOCARPIC.   Once fruiting. Strictly all natural annuals and biennials are monocarpic but the term is usually applied to those plants which take an indefinite period to reach their flowering and seeding age and then die.

OVARY.   In a flowering plant that part of the seed vessel (itself a part of the pistil) which contains the ovules or immature seeds.

PANICLE.   A loose cluster or tuft of flowers borne on several separate branches; an irregularly branched corymb.

PAPPUS.   The variable hairs, bristles or chaffy scales which develop as the seeds of some members of *Compositae* ripen; in certain species they resemble miniature parachutes which assist the seeds to become airborne and to be dispersed by the wind.

PEDICEL. The final flower stalk carrying one flower only.

PEDUNCLE. The stalk of an inflorescence or of a cluster of flowers, but also sometimes used for the stalk of a single flower.

PERIANTH. That part of the flower—petals, sepals or both—which usually creates the display around the plant's reproductive organs. It includes the calyx which is generally green although in some flowers, as clematis, the calyx is the most colorful.

PETAL. One of the separate parts of the inner floral whorl of the perianth. Where, as in lilies, tulips, etc., it is not clear what are petals and what sepals the term 'tepal' is used.

PETIOLE. The foot stalk of a leaf. For the stalk of each leaflet of a pinnate or other compound leaf the term 'petiolule' is used.

PISTIL. A term which includes ovary, style and stigma; the total female organs of a flower.

RACEME. A term used for the arrangement of flowers, like those of Lily of the Valley or Lupin, where a growing main axis bears many separate blossoms the youngest of which is nearest the growing point.

RHIZOME. An underground or surface creeping rootstock or stem, usually growing horizontally.

SCAPE. A stem which grows from the base of the plant bearing a flower or flowers and usually without leaves.

SEPAL. One of the separate leaves which collectively form the calix of the flower. Their main function is to protect the floral parts and they are usually green and leaf-like, but in some plants, as Clematis, true petals are absent and they are the most decorative feature.

SESSILE. Not stalked, or sitting; a term used to refer to leaves and flowers which have no stalk.

SPATHE. The sheath which protects the flowers. In some plants like Narcissus the spathe may be membranous and withers as the flower opens; while in others, as many of the *Araceae,* it is leaf or petal-like and persists.

STAMEN. See ANTHER

STIGMA. The receptive surface of the pistil which receives the pollen grains and through which the pollen tubes pass to the ovary. See PISTIL.

STYLE. That part of the flower which connects the stigma to the ovary. It is not present in all plants.

THYRSE. A flower panicle so shaped that the greatest diameter is half way between the base and the apex.

TUBER. An enlarged part of a root or underground stem.

UMBEL. An inflorescence in which a number of stalked flowers all arise from a central point at the top of the stem. If a number of peduncles spring from a similar point and each peduncle then carries an umbel the inflorescence is known as a compound umbel. It is the umbel characteristic which gives the family *Umbelliferae* its name.